MW01618622

GM
Parts
SIMPSON.
Goodwrench

EARNHARDT
RACING FAMILY ALBUM

LARRY COTHREN
AND THE EDITORS OF *CIRCLE TRACK*
AND *STOCK CAR RACING*

Lowe & B. Hould
Publishers

The edition first published by Lowe & B. Hould Publishers, an imprint of Borders, Inc., 515 East Liberty, Ann Arbor, MI 48104. Lowe & B. Hould Publishers is a trademark of Borders Properties, Inc.

First published by MBI Publishing Company.

On the front cover: Nigel Kinrade

On the title page: *Dale Earnhardt: 23 Years With The Intimidator*: Staff photo
Dale Earnhardt Jr.: Stand Tall in the Shadows of a Legend: Harold Hinson

On the back cover: Harold Hinson

ISBN 0-681-27696-7

Printed in China

CONTENTS

Dale Earnhardt: 23 Years With The Intimidator

Dale Earnhardt Jr.: Standing Tall in the Shadow of a Legend

COMP
76
GOODYEAR
True Value
Winston Cup
MCI
76
3M
Holley
Bud
Gatorade
Mechanix
Bowman
JESEL
Auto Meter
CLEVITE
Nomex
Goody's

CIRCLE TRACK STOCK CAR RACING

DALE EARNHARDT

23 YEARS WITH THE INTIMIDATOR

THE EDITORS OF
CIRCLE TRACK AND *STOCK CAR RACING*

Lowe & B. Hould
Publishers

Harold Hinson

INTRODUCTION

BY DAVID MILLER AND STEVE ZEPEZAUER
Former Editors, *Stock Car Racing* and *Circle Track*

When Dale Earnhardt died at Daytona International Speedway on February 18, 2001, one of the most brilliant and successful careers the world of motorsports has ever known came to an end. The man was, quite simply, a legend, both in his own time and in the hearts and minds of millions of race fans who cheered his every move. He is also a legend for the ages, one of the most enduring of racing's heroes.

At *Stock Car Racing* and *Circle Track* magazines, we were privileged to bear witness to Dale Earnhardt's exploits on the racetrack, his effect on the sport, and his growth from a rowdy young high school dropout to the polished Intimidator of later years. As a tribute to his extraordinary life, we selected photographs and excellent writings from *Stock Car Racing* and *Circle Track* magazines that illustrate Earnhardt's life and racing career.

Our writers, editors, and photographers captured many facets of the man, from the intense competitor to the loving father and grandfather. Through the magazines, we were able to bring fans a picture of Dale Earnhardt that was crisp and clean, yet somehow more full and articulated than the average article tends to be.

You will read about his humble beginnings as the son of another legendary figure—his father, the late Ralph Earnhardt. The elder Earnhardt was a champion racer in his day, and was voted one of NASCAR's 50 Best Drivers on the sanctioning body's 50th Anniversary in 1998. He was also a stern taskmaster, and the program by which young Dale was raised offered lots of hard work for little immediate reward. When he he struck out on his own, the lessons learned from his father gave young Dale the drive and tools necessary to succeed.

You will also read about his successes, which were nearly too numerous to count. As a rags-to-riches story, Earnhardt's has few peers. From the lean, hungry years of his early career, Earnhardt burst onto the NASCAR scene in the late 1970s and began a march that would see him win 76 races, a record-tying seven NASCAR Winston Cup championships and the seat of power in a personal business empire.

As driver of Richard Childress Racing's famed No. 3, Earnhardt won more than $41 million in prize money and six of his seven titles. Shrewd investments and business sense allowed him to earn many times that in the arenas of commerce. He began his own racing team, Dale Earnhardt, Inc., and saw it mature into a race-winning operation by the time of his death. The monument to at least part of Earnhardt's success stands on Coddle Creek Highway outside Mooresville, North Carolina., known to most as the "Garage Mahal."

Dale Earnhardt's life was a patchwork of experience, determination, and a never-abating desire to be the best. Unlike many superstars before him, he was able to take these pieces of life and transfer them to the hearts and minds of millions of fans. He was NASCAR's first true mass-market superstar, and the void left by his death will take years to fill.

Love him or hate him, there are few who would deny that he became the undisputed master of faster and tamed most, if not all, of the tracks that he raced on. As you read the collection of stories in this book, take a moment to reflect on what Dale Earnhardt brought to the sport and to the lives of racing fans everywhere. For those of us who had the privilege of working with him through our magazines, the compiling of these stories was a pleasure.

We have done our best to bring you all the facets of this homegrown superstar, from his simple beginnings to his complex nature and winning ways.

Dale Earnhardt is, and will forever be, a legend. We hope you enjoy our look at the man, the mystique, and the driver.

—Steve Zepezauer, Motorsports Group Director
—David Miller, Motorsports Group Brand Manager

1

EXPLOSIVE

Dale Earnhardt

Stock Car Racing **April 1990**

Maybe you like him, maybe you don't. Maybe you booed him. Even if you have, probably in hushed tones, you've cheered for him, too.

Dale Earnhardt is that kind of person. Those who know him best understand why you might boo him. And they understand, as well, why you cheer for him, too.

What you see is what you get.

He can be tough, cold, friendly, congenial.

No matter what he is, and what he can be, Earnhardt is still the excitement of Winston Cup stock car racing.

What he admires most in his associates is physical toughness and agility and just plain guts. He made a brief statement once while talking about Richard Petty. He was speaking of the toughness of some of the men with whom he is associated, men older than he is; men he may even have looked up to and respected at one time when his career was young.

"Some of these old coots around here are still pretty rank."

I think the statement reveals much of Earnhardt's view of this life and the things of value he finds in it.

I've always felt Earnhardt is probably living in the wrong age, and should have been a cowboy in the trail-driving days. Maybe NASCAR racing is as close as you can come to it these days.

But he's long and lanky, and has a light way of walking about, as if he would be tough in a fistfight.

Living in the century previous to this one, as perhaps he should have, Earnhardt would have looked just right with a gun in a slick holster strapped low on his thigh.

You have to admire Earnhardt, if for no other reason than he is clearly the real article in a sport where too many of the new heroes are worried more about making big bucks than making excitement.

Dale Earnhardt exploded into the 1990s in his black Chevy Lumina. *Mike Slade*

Racing isn't a sideline with Earnhardt. He'll run you on foot to the next lightpole for a dollar, or he might put you in the wall in Turn Four, coming down to the checkered flag with a lot of money on the line. He's unpredictable, but he's also the life of stock car racing. He cherishes this.

Fellow drivers may not vote Earnhardt their most popular pal, but they all carry a great respect for him.

I find comfort, yes, difficult to explain, in knowing that men of Earnhardt's cut are still among us. Men who live by their wits and fortitude, and who believe them that works eats and them that doesn't don't, and who are still too rank and independent to run looking for a federal subsidy every time a hunger pain hits them in the belly.

Earnhardt isn't in that situation now, of course, but he was once upon a time before he became a successful driver. His principles haven't changed. He still believes this way, and if you feel this way, then you and Earnhardt have something in common, no matter whether you boo or cheer him.

But no matter how tough, there are times for remorse. Earnhardt experienced such moments during the latter part of the 1989 season.

It came after his wreck with Ricky Rudd at North Wilkesboro, North Carolina, in October. He showed no remorse or concern for Rudd, but became highly concerned over profanity he used on ESPN immediately following the race.

"What concerned me was all the kids and ladies watching the race on television. I'm sorry for any embarrassment I may have caused. I was upset and said some things I shouldn't have," Earnhardt said.

The 1989 season was an unusual one for the Richard Childress team. Earnhardt, who dominated and won the last race of the year at Atlanta, says 1990 will be better.

"We came up a little short [12 points] of winning the championship," Earnhardt said upon winning the 500-miler at Atlanta, "But I'm already looking forward to 1990 and the Daytona 500. All I need is just a few days off to do some deer hunting and I'll be ready to go."

On the more serious side, and in solitude, Earnhardt talked about what he expected in 1990, and the things that kept him and the Childress team from winning the title in 1989.

"As competitive as we've been over the last few years, I'm looking for another championship in 1990. We will review the 1989 season and study our mistakes, hopefully solve them, and go at it. My theory is if you win races, you'll win championships," Earnhardt says.

He won five races in 1989. Rusty Wallace, the champion, won six. Earnhardt, three-time champion, won five races when he won his first title in 1980, five in 1986 with his second title, and 11 races in 1987.

Earnhardt cites three incidents, claiming a combination of the three cost him the championship in 1989.

He was leading the point race going into Charlotte in early October. A camshaft broke on the Childress-prepared Chevrolet. "That was costly," Earnhardt says. "I was leading by 75 points going into the Charlotte race and came away trailing Rusty by 35 after finishing last in a 42-car field.

Ready to rumble. Earnhardt is strapped in and ready to do battle. *Sam Sharpe*

"It broke our back," he adds, "but I thought we would come back and take the lead.

"Then we go on to North Wilkesboro, and were in a position to take the point lead, but you know what happened there."

Rudd, running second, drove under race-leader Earnhardt going into turn one on the final lap. The two came together, both spinning, with Earnhardt finishing 10th and Rudd ninth. Wallace, placing seventh, gained two points in the season race when he was less than a lap away from losing the lead.

It was on to Rockingham where Earnhardt and Wallace were racing side by side. Wallace hit Earnhardt, spinning him in turn four. Wallace gained 72 points with a second-place finish while Earnhardt came home 20th.

"Those three races were costly for us," Earnhardt says, "I have my own ideas about the two accidents. I can't say either one of them was accidental."

Then, will it be payback time in 1990?

"That would be foolish," Earnhardt says, "but I'll tell you this, it's going to be one heckuva season.

Even Old Ironhead loses it sometimes. *Mike Slade*

"But people just don't understand this part of racing. If I had a vendetta against either driver we're talking about, it would eat me alive. It would take away my competitiveness. Driving a race car, you can't afford to focus your attention on one or two drivers. If I tried to get even after every race, it would be a full-time job.

"I'd have to forget about trying to win, and that is my objective every time I get in a race car."

On the other hand, Earnhardt knows his status in the sport.

"Now, I feel those two guys will sit up and take notice if I'm anywhere close to them. But in 1990 I'll be out there to win races, not beat Rusty or Ricky or knock either out of competition."

DALE EARNHARDT

1990 Season Champ!

2

Circle Track February 1991

Dale Earnhardt claimed his fourth Winston Cup Championship in 1990. *Staff photo*

Earnhardt and the GM Goodwrench/Richard Childress Racing Chevrolet team have added the 1990 Winston Cup championship to an already bejeweled crown of remarkable achievement. Earnhardt's fourth title puts him in a class alone behind seven-time champion Richard Petty, who remains his hero and his role model. In the last 11 seasons, Earnhardt has logged 48 Winston Cup victories. In the seven seasons he has spent with the irrepressible Childress organization, he has captured 39 wins and three championship titles. He leads all of motorsports with more than $12 million in career earnings. All of this glitter and gold makes the missing jewel more conspicuous and enticing. Neither Earnhardt nor his team has won the Daytona 500.

Earnhardt and company are coming off a season "that would have been great even if we had lost the championship," he says. Indeed. He drove the famed "black car" to nine victories (three times more than Ford rivals Mark Martin and Geoff Bodine), had 19 top-five and 23 top-10 finishes, and grossed a record of around $3 million, including the $1 million championship prize and other bonuses.

Nine first-place finishes in 1990 brought Earnhardt and his Richard Childress Racing Chevrolet plenty of attention. *Mike Slade*

He swept at Talladega and Darlington, won the first races at Atlanta and Michigan, and won the second races at Daytona, Richmond, and Phoenix. Completing a comeback that personifies the Childress team, Earnhardt recaptured the points lead with a 51-point gain courtesy of his win at Phoenix. Earnhardt ultimately sealed the championship by 26 points over the pesky Mark Martin/Jack Roush team with a third place in the finale at Atlanta. He ran away with the Winston, sparked Chevrolet to its eighth manufacturer's title, and won his first IROC crown.

"I won't say this title is the sweetest," says Earnhardt, "but it is the most meaningful because it's so hard to win now. I can't say enough about this team for coming back, taking the lead, and beating a team as consistently good as Martin and Roush."

Earnhardt's team overcame consecutive finishes of 30th, 31st, and 34th during the first half of the season and a 25th in the second half. "If we hadn't changed the transmission at Sears Point, a cam at Dover, and put the wheels back on the car at Charlotte [after a mistake in the pits], we never would have won the title," Earnhardt says.

And he might have added that he never drove more effectively. He led 23 races and led the most laps in 10 races, accumulating 165 bonus points to Martin's 80. "Bonus points win championships and getting them was a big part of our strategy," says Earnhardt, who adds that at age 39, "I'll race as long as I'm healthy and competitive, and as long as Childress wants me."

Nobody has to remind Earnhardt, Childress, or the crew that stock car racing's richest and most prestigious individual race victory has eluded them, however. The pain of last year's crushing near miss at the Daytona 500 has been soothed by another championship of 1986 and 1987 vintage. But the tire that denied Earnhardt triumph within a mile of the finish line hangs in the team's shops in Welcome, North Carolina, as a grim reminder.

"The Daytona 500 is the Superbowl of our sport," says Earnhardt. "It just comes at the beginning of the season, instead of the end. To win it is a big goal for us, and we'll go all out as we do every year. We had the biggest buildup and the biggest letdown we've ever had for a race last year, but that's history now. It's a new season and a new race.

"The important thing about winning the championship in 1990, and all the other big things we have won, is not the money, but the sense of accomplishment. That's the way we feel about the Daytona 500." Still, as it turned out in 1990, not winning the showcase event cost the team the $1 million Winston Million bonus.

Earnhardt has a second-, third-, fourth-, and three fifth-place finishes in the Daytona 500, but no cigar. "We'll give Dale the same car that won the other three races at Daytona and Talladega in 1990," says crew chief Kirk Shelmerdine. "Based on last year, we'll probably be the favorite, although it will be hard for us to be better. But we'll have some new stuff. Of course, the other teams will be better too."

If the 1990 season is any indication of the future, the other Winston Cup hopefuls congregating in Daytona better have everything prepared perfectly. The Man in Black, Ironhead, stormed through last season as if possessed and the betting has to be on him again in 1991.

3

EARNHARDT!

The Man Behind the Man in Black

BY BENNY PHILLIPS
Stock Car Racing March 1991

A couple of guys by the names of Gary Rossington and Ronnie Van Zant wrote a song called "Gimme Back My Bullets" about hard times and hard work. Lynyrd Skynrd sang it, and Little Bud Moore, the ex-race driver, remembers how the mornings used to come down when he lived downstairs in a garage apartment and Dale Earnhardt lived upstairs.

"First thing I'd hear every morning was Lynyrd Skynrd," Moore says. "The next thing I heard was Earnhardt's boots hitting the floor."

Earnhardt was living the words to the song at that time in his life. Today, he probably still enjoys slipping "Bullets" into the cassette player, kicking back, and remembering the days when he was pulling himself up and getting ahead when the odds were steeper than a Daytona turn.

And he no doubt realizes you can tumble from a Daytona turn if you don't keep your foot in it. So, his life—or the priority in his life—is driving a black Chevrolet Lumina stock car. He seldom looks back and is never a welcome sight in a rearview mirror. If you see him coming this lap, chances are you will see him going the next.

Right on past, right on down the line. He does some pretty dangerous things, on and off the track. But the occupation he has chosen has carried him to undreamed-of heights and wealth—and that is what is important to today's Earnhardt.

In 1990, he captured his fourth title in stock car racing's premiere series, and only Richard Petty, with seven, has won more. "We're young enough to maybe catch Petty," he says with a giant glint in his eye. But Earnhardt will be 38 longer than Jack Benny was 39.

Petty's feat of winning seven NASCAR titles once seemed as unassailable as Joe DiMaggio's 56-game hitting streak. But given the combination of Earnhardt's temperament and driving ability and the crack team that Richard Childress has placed behind him, uneasy lies the crown on King Richard's head.

Harold Hinson

If ever there was a photo that captures Dale Earnhardt as he's seen by the men he beats, this is it! *Staff photo*

It is ironic to a degree, because no two competitors could be as different as Earnhardt and Petty. Petty is known almost as much for his easy manner with race fans and friendly practices with the press as for his ability on the track. During his prime, there was never any doubt about Petty being the most popular driver.

Earnhardt has never won this award. It went to Darrell Waltrip in 1990.

A panel of press members also voted Al Unser Jr.—not Earnhardt—"Driver of the Year" in 1990.

Earnhardt has to force himself to do the things off the track that are demanded of the modern-day driver. Obviously, he does a better job on the track.

"The most fun I have is when I'm in the race car competing," he often says.

Earnhardt is a competitor in any endeavor. He doesn't want to be second in anything he does, including a friendly game of pool. He is intense.

Always charging to the front of the pack, Earnhardt built his reputation on being intimidating. *Staff photo*

In 1989, Earnhardt came in second to Rusty Wallace in the point standings, beaten by a mere 12 marks out of the many thousands available in the 29-race season.

He remembers sitting in a deer stand afterward, brooding about it, remembering the points he literally had thrown away.

"The more I thought about how we had been beaten for the title the more it bothered me," he recalls, "because we had given the championship away with our performances in two key races—North Wilkesboro and Rockingham."

The race at North Wilkesboro was probably the most galling because he was leading on the final lap when Ricky Rudd tried to pass him. Had he let Rudd by and settled for second, Earnhardt would now have five championships.

Instead he tried to hold Rudd off, both spun, and Earnhardt finished 10th. He should have remembered that he and Rudd had not gotten along that well previously. He should have known, too, that Rudd is not one to back off, and that he had less to lose in a last-lap tangle. But Earnhardt believes he shouldn't back off from anyone.

The eighth-grade dropout has had his share of confrontations with other drivers because of this theory. "We all set limits on how far we'll go, the things we'll do to win a race," says fellow driver Geoff Bodine. "We all don't agree on those limits. Dale and I, we've had different values and they've clashed at times. Who's to say who's right and who's wrong?"

"He's as determined to win as he was the first time he ever sat in the car in 1981," car owner Richard Childress says of Earnhardt. "He still has that drive to win. He doesn't want to lose at

Richard and Judy Childress with Teresa and Dale Earnhardt on stage at the Waldorf-Astoria to celebrate the 1990 Winston Cup Championship. *Staff photo*

anything, from fishing to deer hunting to racing to whatever.

"But he's different, too. He's smoother but more aggressive in his driving style than he used to be. Watch him race. The chances he might have taken in 1981, he can take the same chances now but he's a lot smoother taking them. He looks ahead, thinks things out. That comes with maturity."

Earnhardt has raised hell and eyebrows since he became NASCAR's 1979 Rookie of the Year. He has gone through fleets of race cars, a couple of marriages, and has yet to face the foe to whom he'd knuckle under.

He has won 49 major-league races and is considered by some the best in his line of work. But there are many more than a few who don't like him or cotton to his style. They never did and never will.

"He's good, " Richard Petty says. "Real good. But he's not really my kind of race driver. I used to beat people but I never ran 'em in the ground."

But what you get today is not what he was back when he was making time between his home in Kannapolis, North Carolina and the Great Dane Trailer Co. in Charlotte, where riveting kept him in gas money that got him to the track and back. Those were hard times and lean times for a kid whose fix was the scent of gasoline fumes and burning tire rubber. But that's the way it had been since he first learned to hand his daddy a cold swig of water and the right wrench on a dusty dirt track Saturday night.

"I learned a lot from Dad," Earnhardt said. "I watched him be patient and work people to the end of a race, and then win. But I never tried to fashion myself after anybody. I just try to be myself, not my dad, not Richard Petty.

"From the time I started I didn't have a goal in mind. I never thought I'd be a champion, but I knew I had the chance, and every chance was an opportunity I took advantage of, from the time

Quick pit action made for great track position for Earnhardt and Richard Childress Racing. *Mike Slade*

they put me in a car and I won the rookie title."

His greatest regret, he said, is that he didn't make his mark before his father, Ralph, died—of a heart attack while working on his race car in a garage beside their home.

He regrets, too, he said, that during two previous marriages two sons and a daughter were too often left without a father's hand to hold because his were too often wrapped around a torque wrench or a steering wheel.

Today, Teresa and Dale Earnhardt seldom are far apart. And after he wipes away the grime of the 500 miles, among the first to share his good or bad times are his wife and young daughter, Taylor Nicole.

"I'll never do it again, but at one time I gave up too much to racing," he says.

Teresa seems as gritty as the guy she married. She's a friend and personal business manager. Very often she is seen atop infield towers, herself in a corner, against a rail, feisty little fists by clenched eyes fixed on that hunk of cold black speed, the one with the number three that is so often number one.

And today there is success. There's a sprawling lake front home at Lake Norman, the farm in the country, the Chevrolet dealership.

Still, more than ever before, there are those tugging at his shirttail, wanting one small piece of his hide. And he has more hangers-on than anybody ever had in racing, the crowd sometimes reaching the size of the bloodsuckers who trail a heavyweight boxing champion.

"It's the competitiveness of the sport," Earnhardt says. "You learn to handle what you can. But there's a lot more to it than just racing.

"It's difficult. I have so much going on personal-wise and sponsor-wise that it's difficult."

Earnhardt says he can gauge his time by the time he has for deer hunting. "I used to hunt about three months a year; now it's down to about a month."

Dale Earnhardt is joined by daughter Taylor Nicole and wife Teresa in the winners circle after his 1991 Busch Clash win at Daytona. *Staff photo*

Earnhardt's office is decorated with posters from his career. *Staff photo*

Earnhardt has traded sheet-metal punches with the best of the rest; he's been fined, threatened, cursed, and booed. And he believes he has reached the pinnacle of the sport because he has stuck to his guns.

Whenever NASCAR sat on him because of this aggressive driving style, he hardly flinched. The sanctioning body had to do what it had to do, he said, but so did Dale Earnhardt.

"I can't let them set the guidelines for my racing style," he says. "I have to set those guidelines. I'm not going to try and hurt anybody. I've made mistakes. I've been wrong. But I've never set out to wreck or hurt anybody.

"You roll with the punches and you have to be consistent. When I got into it with [Bill] Elliott at Charlotte in 1987, I stuck with my story and I got fined. Darrell [Waltrip at Richmond in 1986], I told them it was my fault and I got fined. I always stated my case. I've always tried to be the same.

"But I've mellowed, too. I've watched them all, and I admire Richard Petty, the way he gets out of a car, puts on that hat, and smiles."

Earnhardt does have a growing respect for his fans. "Used to be, you didn't think much about the fan. He bought a ticket and he was up there sitting in the stands. But I went to a deal in Greenville, South Carolina, after Darlington on Labor Day, and there were about 8,000 people there. There might have been one out of a hundred who didn't have something on that said Earnhardt. My dad drove a Chevy down there in the 1950s and won the track championship, and I saw people there who not only remembered him, but who remembered me. An old friend of Dad's walked up and said, 'I can't believe it's you. You used to sit there on the back of a truck when you weren't nothing.' "

Some say Earnhardt has mellowed in the last year or two. They are usually quick to add it probably has made him a better driver.

And Mark Martin, who battled Earnhardt down to the final lap of the final race of the season before finishing second by 26 points in 1990, says, "It was all good, clean racing. There was never a negative word, just a good, clean battle that the fans loved.

"I see people who have problems with Dale on a regular basis. I know it goes back to the things that have happened. Once in a while I have to take a little from Dale to get by and sometimes I have to give a little, too. I have to make judgments on when it's time to give him some room and when I have to take some room.

"But he has never, ever blocked me on the race track, other than to run his line or my line and forced me to take another one.

"He always gives me a line. It may not be the choice one. He may have watched me and said to himself, 'Mark wants to run high, so I'm going to get up here and he'll have to pass me on the low side. If he can, he can have it.' "

Martin thinks Earnhardt's edge is "he's never stopped charging hard, but he learned how not to get in trouble. He was stubborn enough not to give up charging but dedicated and smart enough to finish and win.

"You've got to be able to charge and bring 'em home, and Dale has learned to do both. Not everybody can do that."

Earnhardt had an awesome season in his chase with Martin for the title. He won nine of the 29 races and the International Race of Champions series. His earnings of nearly $3 million was a record for any race driver in a season and he came within an eyelash of winning nearly a million more.

If he hadn't cut a tire on the last lap of the Daytona 500 he would have won another $1 million bonus for winning three of the four races that comprise the Winston Million. As it was, he pocketed an extra $100,000 for winning two of the four.

"I feel that we got beat by the best race driver in the business behind the wheel of the best car he ever sat down in," Martin says.

"It was a good, competitive battle," Earnhardt says. "We weren't eating buddies, but it was a good clean battle, Ford and Chevrolet.

"I put a little black paint on his car, he put a little red paint on mine, and there were times when neither of us knew how far to take it. But I guess the difference was we both figured as long as both were able to come out of the corner, we were OK."

"He's as determined to win as he was the first time he ever sat in the car in 1981," car owner Richard Childress says of Earnhardt. "He still has that drive to win. He doesn't want to lose at anything, from fishing to deer hunting to racing to whatever."

4

EARNHARDT

On Racing and Education

BY BRUCE MARTIN
Stock Car Racing July 1991

Dale Earhardt enjoys riches and fame that few humans could hope to achieve in a lifetime. At 38, Earnhardt has won four Winston Cup championships, 48 races, and over $13 million. He has also won fame and adulation that can't be measured.

But there is one thing that is missing from Earnhardt's life, and he believes he needs to be a role model to make sure others don't make the same mistake he did. Earnhardt does not have a high school diploma.

If Dale Earnhardt has any advice for a teen who aspires to become a race driver, it's finish school first because it will greatly help later in racing. "You work hard and things come to you, but still, there are days through my championships and my wins that I wished I had a better education and made better decisions," Earnhardt says. "I've had to rely on a lot of common sense and good advice from my wife, Teresa.

"I think an education would have helped me a lot. It would have helped me from the sponsor side of it and the fan side of it."

When Earnhardt was a youngster, he quit school after eighth grade against his father's wishes. Young

For Dale Earnhardt, family and education are as important as his racing. *Mike Slade*

Dale's sons Dale Jr. *(left)* and Kerry *(right)* have followed in their father's footsteps. *Harold Hinson*

Dale wanted to help his father, the legendary NMPH Hall of Fame driver Ralph Earnhardt, and learn about racing. "I was young and dumb and I should have listened to my father," Earnhardt says now. "I've made sure my kids don't make the same mistake I did."

Earnhardt sent two of his kids, Kelley and Dale Jr., to private schools until the 11th grade. Then, he allowed them to go to public school.

Kelley has since graduated and is now a student at the University of North Carolina-Wilmington. Earnhardt's son is a junior at Mooresville High School in Mooresville, North Carolina.

"I wanted to make sure they had a good high school education and then have the opportunity to go to college," says Earnhardt. "I'm able to do that with my racing and my profession. If I hadn't been lucky, they might not have been able to go to college."

As Winston Cup racing becomes increasingly competitive, and sponsors demand more and more from their drivers, Earnhardt believes a good education is more important than ever. "Racing

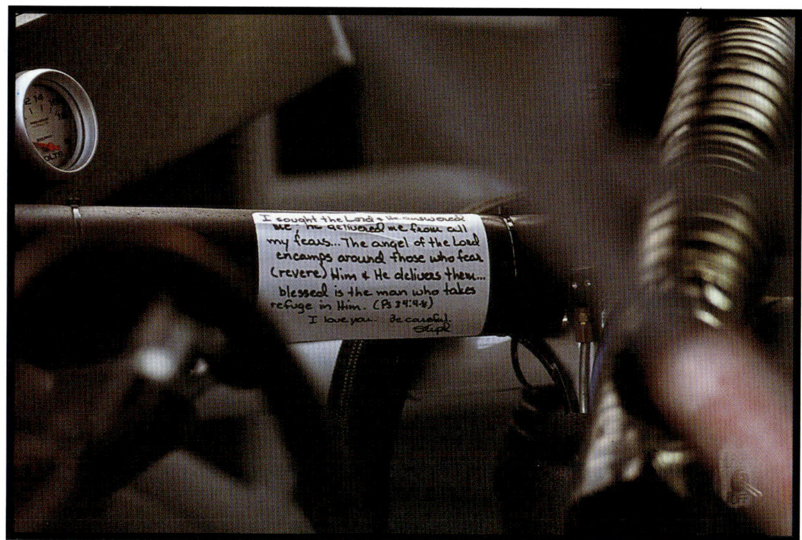

A scripture to race with. *Nigel Kinrade*

has gotten so competitive with the computers and the technological advancement, it's crucial to thoroughly understand what is involved," he says. "When they send a car to the wind tunnel now, they do things on the race car that I don't understand. I understand what it does, but I can't come up with the equation to tell you what they should do.

"I know an education would help me in that regard."

That is the warning that Earnhardt gives to any young high schooler who is beginning a racing career. While they think they may have the know-how to put a competitive car on the short tracks of America, it's a much different story in NASCAR, where the art of stock car racing has become a science.

An understanding of engineering and aerodynamics can help on the track, with public speaking and a sharp business acumen key ingredients off the track.

"If they don't have an education in this day and time, they will be dealing with people who are so smart, they will be intimidated," Earnhardt says. "The guys around them are going to be talking on levels so much above them. The crew members are at the point where you need an engineering degree to understand them."

Earnhardt has reached the top of Winston Cup racing because he can drive a stock car better than anyone else. But he credits his wife in helping him understand what is important off the track. "We got married in 1982 and it's been just a great relationship, " Earnhardt says. "She has tried to better our lives in business and racing. She helps us with contracts and other company involvement. She does a lot of things with endorsements and sponsors. She comes up with good agreements between companies, sponsors, and myself. She has done a good job with that.

"She does a lot of things I don't see. There are times you think it's not too important to go over here and speak, or show up at this function. She will show me the reasons why it might be a good idea to do it."

It's obvious that Earnhardt has become more polished since he was named Winston Cup Rookie of the Year in 1979. "I really wasn't mature enough," he says. "I was a happy-go-lucky guy. I won the title in 1980 and I was in it for five years before I won my second title. I respect it and enjoy it more because I know what is at stake.

"I didn't realize what racing and my profession were really about. I was no longer just a part-time racer, it was a business—a big business. I've learned to work with it, grow with it, love it, and respect it.

"I would do anything to help the sport grow. I've worked to be a better champion. Our sport has grown so much in the last several years, and it's a huge change since I started in 1979.

"You need to learn," Earnhardt continues. "I think I have learned."

Undoubtedly, Earnhardt has learned the value of patience, both as a race driver on the track, and as a public figure off the track. "I think patience has helped me with the driving part of my profession and patience with the press and the fans," Earnhardt says. "Richard Petty is the King. He has won seven championships, but if you sit and watch the man day in and day out, he smiles and takes a minute for the fans.

"Even when he walks past fans going to his car, knowing he doesn't have time to stop, he's still good to them. He smiles and waves and gets in his car. When he has a good day, he talks to the press. When he has bad day, he talks to the press. He always gives the press time and he's been a great champion.

"If you watch him and people like him, you'll learn to be a good champion, too."

Only Petty's seven titles exceed Earnhardt's four. With each title, Earnhardt has learned what it takes to be a champion, and how to act like a champion. He believes each championship is special in its own way. "Winning with Richard Childress for the third, and my fourth title is part of why it is special," he says. "We had a tough battle with Mark Martin and the Roush team. Getting beat in 1989 by Rusty Wallace is part of why it's special.

"I think as you get older, and you've won races and championships, they come harder and you know you have less chance to win a championship. I'm 38 years old and hopefully, I can drive until I'm 50 and drive healthy and competitive. But the day will come when you're not as competitive as you should be and you'll be slowing down. Your chances of winning a championship become fewer and fewer.

"It feels good to still be competitive and dominant," Earnhardt continues. "This last title says the Richard Childress team is the best of the last several years. This is five years in a row that we have been a factor in the championship battle.

"To win three of those five years is pretty impressive, but to win the championship four times in the last 10 years is an accomplishment I'm proud of. It makes me feel very good.

"I'm not in any way comparing myself to Richard Petty, because he is still the King, but to be second in line is pretty impressive. It helps put our team in a class all by itself."

The ability to learn from past mistakes is one of the reasons Earnhardt has been placed in a class by himself. Now that he has achieved the status of a role model, he urges youngsters interested in racing to stay in school and get a good education before starting a racing career.

5

DALE EARNHARDT

The New King?

Circle Track December 1991

Staff photo

Dale Earnhardt has picked up the moniker "the Dominator" for obvious reasons. With four Winston Cup championships, 51 race wins (at press time), and a reputation for always being near the front of the lead pack, there have been whispers (ever so subtle, mind you) that the time may have come to transfer the title of King from the red and blue No. 43 car to the Man in Black.

There are a lot of people who would argue that it is blasphemous to call anyone but Richard Petty "the King." They might be right. Earnhardt's racing career has matured in a time when the competitor is tougher and more intense due to a more competitive field of cars; however, Petty's career still includes some incredible feats. While Earnhardt was impressive in 1987 with 10 wins, Petty won 27 victories in 1967 including an amazing string of 10 in a row.

While Earnhardt might catch up to Petty's seven national championship titles by the time his career winds down, it's doubtful that the under-30 race schedule of the modern-era Winston Cup series will allow him to get close to Petty's 200-victory milestone.

Ask most any spectator who the race favorite is today, however, and the answer will usually be "the Dominator." No matter how you feel about the subject, it is a good idea to remember that Petty wasn't always the King. He started out as "the Randleman Rocket." Is a name change on order for Earnhardt?

Is the "Man in Black" poised to take over the throne from "The King"? *Staff photo*

There are a lot of people who would argue that it is blasphemous to call anyone but Richard Petty "the King."

6

PRESENTING THE WINNER

The Pressure's Already on Dale Earnhardt for Three in a Row . . . and More

BY BOB MYERS
Circle Track March 1992

While it may seem that Dale Earnhardt has won it all more than once, there are at least three more jewels that the irrepressible five-time Winston Cup champion wants to affix to his sparkling crown before his sensational driving career ends. Having added the 1991 championship to a growing list of milestones and accolades, Earnhardt has set his sights on seven titles, a third straight, and victory in the Super-Bowlish Daytona 500 by STP—in reverse order.

A seventh crown would tie all-time leader Richard Petty. Three in a row would match Cale Yarborough's feat of 1976-1978 in Junior Johnson's hardware, a milestone that Earnhardt was denied when he finished third in 1988. Of all that Earnhardt and the crack Richard Childress–owned RCR Enterprises/GM Goodwrench Chevrolet team have mastered, it is incredulous that they haven't won a Daytona 500 yet. The sport's richest and most prestigious showpiece is their nemesis, a source of frustration and bitter disappointment, none greater than in 1990 when victory was shredded by a cut tire in the last mile.

Earnhardt, 40, an amazing talent who has raced hard from the valleys of hope to the summit of superstardom, from pennies to more than $16 million in motor sports earnings, making the black car a symbol of notoriety second only to Petty's famed No. 43 believes the milestones are within reach. "Richard Petty is the King and always will be," says Earnhardt who respects and admires the winner of an untouchable 200 races, 7 titles, and 7 Daytona 500s. "But I really would like to win seven championships. Now that we've won five, I can see that's possible. Harry [Gant] is winning big at 51, so I feel I've got 10 to 11 more years. If I can stay healthy and Childress doesn't fall off a mountain on a hunting trip, I think we'll be in good shape for another decade." That is, in spite of vicious rumors, Earnhardt says, aimed at splitting him and Childress.

"It would be nice to tie Cale's record and we're going to conquer that Daytona 500 before I get too old." Earnhardt continues, "But if we are to accomplish both in '92 we've

It may not have been a pretty victory, but Dale Earnhardt won the championship nonetheless. The ability to cool it when required demands a fair amount of driver maturity. *Staff photo*

got to have a better season than 1991. We won the title and we'll take it, but we didn't impress anybody with the way we raced. We were way ahead in points, partly because the others were off when we were or arrived too late as contenders for the title."

The first order of business for the championship-proven Goodwrench bunch is to win the Daytona 500, set for February 16 at Daytona International Speedway. In 14 starts in the premier race since his first in 1979, Earnhardt has 7 top-five finishes, highlighted by a second in 1984, third in 1989, and fifths in 1990 and 1991. "It's our biggest race. There's more emphasis on it than any other and we've come so close, we're driven to win," says Earnhardt. "With all our championships and wins [52 for him], not winning the Daytona 500 is like wearing a tux with a pair of brogans. We're going loaded for bear even though we don't know how engine rule changes and some cutbacks by General Motors will affect us and the rest of the field. And if we can win, maybe we can buck that other trend and go on to the 1992 championship." He alludes to the fact that no one has won the Daytona 500 and the championship in the same year since Richard Petty in 1979.

The sport's showcase event is certainly top priority with the Childress team. "We want the race badly," says veteran crewman David Smith. "I've told Richard not to get rid of me, even if I'm 80, until we win a Daytona 500.

"We've been so close and really haven't done anything wrong. It's just been a matter of circumstances. We've had the fastest car and the best car. Now we want to have the car that gets there first, by luck or whatever."

"The Daytona 500 is the goal of every man in this shop," says foreman Cecil Gordon, who has not been a part of a winning team in the 500 during his long career as an independent driver and six years with Childress.

"We were a lot better in the Daytona 500 in 1990 than we were last year," adds crew chief Kirk Shelmerdine, "and we aim to regain that form. We are not completely proud of our '91 season. Some races we were flat mediocre. We've got to reestablish ourselves as the dominant car and team besides being the champion."

As early as last November, the team's new Daytona Lumina was almost finished and Childress was working long hours charting engine combinations for engine chief Eddie Lanier and crew to build for tests. The new car relegates to backup status the trusty Lumina in which Earnhardt finished fifth twice in the Daytona 500, won three of the four restrictor-plate races in 1990, and the DieHard 500 at Talladega in 1991. Is the new car a wishful change of luck? "Naw, we build a new speedway car about every two years," says Shelmerdine.

Lanier says new Chevrolet cylinder heads approved by NASCAR should be an improvement, "but they're basically an unknown until they are tested and raced," he says. "With the 7/8-inch restrictor plate, we're probably looking at a gain of 4 or 5 horsepower over last February, up to 420–425, and probably an increase in speed of 2 mph, perhaps to 197, with engine restrictions working in tandem with NASCAR's rear spoiler rules and other aerodynamics aids."

Crew chief Kirk Shelmerdine *(left)* and owner Richard Childress *(right)* hope to win more championships with Earnhardt driving. The combination still carries a mystique that others aim for as a goal. *Nigel Kinrade*

By almost everybody's yardstick, Earnhardt had a great 1991, but not by his and the team's lofty standards and expectations. He wasn't "the Dominator" and rarely was "the Intimidator," but he certainly was "the Terminator." Earnhardt's four victories (compared with nine in 1990) were the fewest since three in 1988 and his points average of 147.428 per race was the second lowest by a champion since the present points system was adopted in 1975. Rusty Wallace beat Earnhardt in 1989 with an average of 144 points. In the title year of 1986, Earnhardt averaged a whopping 161.931 points. The highest average since 1975 was Yarborough's 166.666 in 1977.

Earnhardt clinched the 1991 title—his team's fourth—by merely starting the season-ending Hardee's 500, finishing fifth at Atlanta Motor Speedway, and beating Ricky Rudd and the Hendrick Motorsports/Tide Chevrolet outfit by a margin of 195 points. However, Rudd won only one race and had 9 top fives, compared with 14 by the champion. Late charges by third-place Davey Allison and the Robert Yates Racing/Havoline Ford team and fourth-place Gant in Leo Jackson's Skoal Bandit Oldsmobiles, each with a series high of 5 victories, fell short. The second through seventh finishers in the 1991 points standings—Chevrolet's Ernie Irvan was fifth, Ford's Mark Martin was sixth, and Ford's Sterling Marlin seventh—have never won a Winston Cup championship. Allison ranked 21st in points, 239 behind Earnhardt, after the third race of the season and earned more points than the champion until the final race when he lost second place to Rudd by 4 points. Other categories Earnhardt habitually leads and didn't in 1991 include most laps, miles, races, and times led. Earnhardt also failed to win a pole (4 in 1990), a shortfall that excludes the defending champion from the February 8 Busch Clash, while Martin led the series with 5 and Allison had 3 of a season-leading 18 by Ford drivers.

Still, though Earnhardt won only 2 of the final 13 races, Chevrolet withstood an onslaught by Ford to covet its unprecedented ninth straight Manufacturer's Championship (and 14th in 16 seasons) by three points (184 to 181), on the strength of Irvan's second place in the Morgan-McClure/Kodak Lumina at Atlanta. Oldsmobile was third with 136 points, followed by Pontiac (two wins each by driver/owners Wallace and Darrell Waltrip and a solo by Kyle Petty) with 115 and winless Buick with 80. And Chevrolet once again beat archrival Ford in victories, 11 to 10, though Fords swept the last four races. There were a record-tying 14 different winners in 1991, but three drivers who won in 1990—Brett Bodine, Derrike Cope, and Morgan Shepherd—failed to repeat. Dale Jarrett, driving the Wood Brothers/CITGO Ford, was the only first-time Winston Cup winner among such hopefuls as Michael Waltrip, Sterling Marlin, Rick Mast, Hut Stricklin, Jimmy Spencer, and Bobby Hamilton. However, Hamilton edged Ted Musgrave as Rookie of the Year by 17 points. Hamilton posted four top 10s and finished 22nd in points in George Bradshaw's Tri-Star Racing/Country Time Oldsmobiles to win the honor, worth more than $50,000 in bonuses.

Pit stops have always been extremely important in circle track racing, especially when the race is 500 miles long. Clichés like "clockwork" and "incredible" don't come close to describing the impact the Goodwrench crew has had on the team's success. *Staff photo*

In a strategic move to protect his team's points lead, Childress resorted to a conservative late-season game plan that reduced Earnhardt to something of a "Timidator."

"Our championship wasn't pretty but we won it the only way I figured we could," says Childress. "I was convinced we couldn't race the Fords head-on without blowing up, and considering Rudd's consistency, we backed off, particularly at Rockingham and Phoenix, to protect our lead. We had raced our rear ends off for 25 races to get a lead and found ourselves in a gunfight with a pocket knife. I wasn't about to give the title away in the last few races. People say it's worth $1 million but it's closer to $2 million. We'd already tried to give it to Rudd and he wouldn't take it. After we went for it [in the Mello Yellow 500] at Charlotte and broke while leading, we switched to engine stuff that we used in late 1990. It was 30 horsepower below what Ford had, but we knew it was reliable. We got seventh at Rockingham and ninth at Phoenix, where we would have had fifth if we hadn't missed the setup. We essentially clinched the title in those two races and avoided the possibility and the risk of going to Atlanta with a slim lead and having to race Rudd and the Fords hard."

Here's an uncharacteristic sight of Earnhardt. No one can accuse him of having all the luck go his way. *Staff photo*

A tattered Chevrolet flag flies over the RCR Enterprises shops in Welcome,

We didn't get to see the untouchable runs through the pack in 1991, but the team might get back into polished shape with a few changes in the engine department. *Mike Slade*

North Carolina. "That's what Ford did to us for two-thirds of the season," says Childress, who states that Fords gained the advantage by using cylinder heads under a different part number, which weren't supposed to be approved until 1992 and with an additional rear deck height of 1 inch. "I'm not knocking NASCAR," he emphasizes. "They did a good job with the rules, but sometimes one brand gets a little advantage, and Fords had the best of both worlds—horsepower and downforce—during the latter half when they won all of their races. Chevrolet's new cylinder heads might not catch up to Ford in 1992, but they will help and certainly will cut operating costs."

In spite of vicious and nasty rumors (all flatly denied) that claim Earnhardt's wife, Teresa, "caught" him with Childress' married daughter, Earnhardt is under contract to drive for Childress for two more years and indicates that the two will be together for many years to come. "I can't imagine who started such trash," says Earnhardt, who appeared with Childress before the media at Atlanta to set the record straight. "I guess if they can't beat us on the track, even in an off year, somebody figures they can split us up with rumors. Well, the rumors didn't work, except in reverse to make us even stronger. The stuff is garbage and we want to set the record straight if it needed to be."

Even when he isn't winning races, the Dominator has no trouble attracting media attention. The "one race at a time" plan might have released some of the pressure. *Staff photo*

"It's unfortunate that there are people who want to destroy reputations and disrupt a team," says Childress. "It's particularly sad when innocent people are involved. My daughter has been torn by the rumors, none of which are true. I told her that people shoot from all angles in this business. Dale and I are good friends and we have a lot of respect for each other and our families. Who knows, maybe in time he will want to operate his own team and if he does, I'll be the first to help him get started. We don't always see eye-to-eye, but we always talk out our differences. He didn't like the conservative game plan, for instance, but he had the maturity to know what we had to do to win the championship. He did it and that's a compliment to him."

Childress paused, with thoughts of 1992. "It's going to be a humdinger, trying to win three straight titles," he says. "Strong championship contenders emerged in 1991—Allison, Rudd, Irvan, Gant—who have not won a title, not to mention Roush Racing and those who have won in the past. We've been fortunate to have finished no worse than third in the title quest since 1985, but every year the championship gets harder to win. The competition might beat us, but we don't intend to get beat by rumors."

Seven championships for Earnhardt? That's all right with the King. "He's the only one on the horizon who has a chance to do it," says Petty, "and I'd say that if he had won only three at this point. More power to him. That team has the combination—good owner, crew, motor man, and luck—that translates to championships."

7

EARNHARDT by Numbers

Stock Car Racing March 1994

Mike Slade

Six Winston Cup championships . . . 59 Winston Cup wins . . . more than $19,000,000 in career earnings. Those are the figures you're familiar with. But what's the story behind the stats?

15,000

The number of fans on hand when Earnhardt wrapped up his first Winston Cup championship at Ontario, California, in November of 1980.

Dale Earnhardt's rookie season in Winston Cup, 1979, coincided with the arrival of live television. Throughout the 1980s, ever-increasing TV time pushed stock car racing—and Earnhardt along with it—to levels of popularity few could have foreseen at the dawn of the decade. "Our sport has grown so much in the last several years," Earnhardt told *SCR* in 1991, "and it's a huge change since I started in 1979."

No number better signifies that change than the attendance figure for Ontario Motor Speedway's 1980 season finale. Can you imagine what would happen now if the Winston Cup schedule concluded at a Southern California superspeedway?

Thirteen years and five championships later, Earnhardt has changed almost as much

First championship: 1980. *Staff photo*

as the sport. "In many ways, I really wasn't mature enough," he's said of how he handled himself during his years in the Rod Osterlund car.

There's ample evidence of that. Early on Earnhardt could be crude, and he was confident to the point of cockiness. "Stick with me," he allegedly told crew chief Jake Elder, "and we'll be wearing diamonds as big as horse chips."

Yet it was those very qualities—the bravado and the earthy language, which were reflected in his driving style—that helped make Earnhardt and his sport such a hit in the 1980s. As Humpy Wheeler put it: "Dale Earnhardt has produced tremendous drama in stock car racing. He's unpredictable, he's intimidating, he does things on the track that a lot of other drivers won't do and gets away with it."

"In 1980, there wasn't a lot of pressure on me. Everyone was asking, 'Who is this kid? Cale's going to win it.' I lucked up and won it. It wasn't a [fluke]."

10,000 Feet

The altitude at which Earnhardt and Richard Childress went elk hunting last fall following the Rockingham race.

Does this shared interest in hunting have anything to do with the success and longevity of their relationship? Almost certainly. "Richard Childress and I were talking on that mountain," Earnhardt said after the trip. "We were up about 10,000 feet on a big mountain and wondering how many drivers and car owners were doing something like this together. I think our relationship is going to help the length of my career. If I see a fault in me or Richard sees it, we'll talk about it and try to do something."

The fact that these guys go to such extremes in the name of recreation also reveals something about their character. They play as hard as they work—and since they make their living in professional sports, that's a healthy attribute. As Earnhardt put it at Rockingham: "Right now, I'm going to fly to New Mexico and get after an elk tomorrow wide open."

Sometimes, you just need to leave the job behind. Recognizing this, Childress shut his entire shop down for a week midway through the 1993 season. "When you drag out hours, [employees] tend to get burned out," he explained. "We've found out in the past when you give people time off, it helps out a lot."

$3,000

The amount Earnhardt was fined following a late-race crash at Richmond in February 1986.

The Richmond incident, along with the "Pass in the Grass" (a misnomer—actually, Earnhardt was maintaining the lead, not passing) at Charlotte in 1987 probably did more than anything to solidify Earnhardt's reputation as an aggressive—at times over-aggressive—driver. He was a charger who offered no apologies. Sometimes it worked, as at Charlotte. Sometimes it didn't, as at Richmond.

The situation was this: With three laps to go, Darrell Waltrip passed Earnhardt for the lead. As Earnhardt tried to take the lead back, he caught Waltrip in the right rear and turned him toward the wall.

The resulting crash involved not only Waltrip and Earnhardt, but Joe Ruttman and Bobby Allison. Kyle Petty, the last driver on the lead lap, went on to his first Winston Cup win.

After reviewing the incident, NASCAR determined that Earnhardt was guilty of "reckless driving" and fined him $5,000, which was reduced to $3,000 after an appeal.

Earnhardt was unrepentant. "If I was trying to wreck Darrell," he said, "I wouldn't have wrecked myself, that's for damn sure."

Career short track victories: 23. *Nigel Kinrade*

2001

The year some say Earnhardt turns 50.

Officially, his birthdate is listed as April 29, 1952, but there are those who insist he was actually born in 1951—or earlier. As *SCR*'s Benny Phillips wrote in 1991, "Earnhardt will be 38 longer than Jack Benny was 39."

Earnhardt's true age is irrelevant. There's still plenty of fight in him. So how much longer does he plan to race? "That's something you can't answer," he said late last season. "I'm 42. I feel good. I feel healthy. I'm racing good. I'm driving good. I feel like my reflexes are sharp. I make good decisions on the race track. I don't crash too many times. If I just keep the pace up, I think I'll be OK."

OK enough, perhaps, to break Richard Petty's record of seven Winston Cup championships.

500

The estimated number of Ralph Earnhardt's short-track wins.

Earnhardt's father, whose death in 1973 Dale once described as "the biggest shock of my life," had a greater influence on Earnhardt than anyone else. "Racing and the outdoors were my dad's life," Earnhardt told *SCR* in 1979. "It became mine at an early age. I can remember sitting in school, counting the seconds ticking off the clock until class was over and I could run home and help him in the shop."

489

The point differential between Earnhardt and runner-up Bill Elliott when Earnhardt won the Winston Cup in 1987.

"I have been wrong. I admit it when I am. I was wrong at Richmond when I crashed Darrell, but I wasn't wrong in the Winston. I wasn't going to let Bill Elliott beat on my rear bumper, and I won't the next time, either."

Career superspeedway victories: 36. *Nigel Kinrade*

Hands down, Earnhardt's performance that year was the most dominant of the last decade. During that span, Earnhardt, Bill Elliott, and Rusty Wallace have been the only drivers to win 10 or more races in a season. Only Earnhardt managed to win the championship the same year, however. His 11 wins in 1987 included four in a row, the same as Harry Gant in 1991, Bill Elliott in 1992, and Mark Martin in 1993. But Earnhardt also threw in mini-win streaks of two and three for good measure. In fact, that spring, his two-race and four-race streaks were separated only by an Atlanta event at which he won the pole and led 196 laps, only to fall off the pace with a faulty master switch. Had he held on, he would have won seven straight.

The point race that year had all the drama of a rain delay. Besides winning races, Earnhardt finished races, recording 24 top-10s in 29 starts. His only two DNFs came late in the season during what amounted to an extended R&D session for the Childress team. When Earnhardt scored his final victory of the season at Richmond on September 13, capping a three-race win streak, his

Championships with Richard Childress: 5.
Mike Slade

point lead peaked at 608. That meant he could've skipped the next three starts altogether and still been on top. As it was, he wrapped up the title at Rockingham.

"I never thought I would win back-to-back Winston Cups," Earnhardt said when it was over. "Your luck has to be enormous to stay out of trouble on the track, have the team perform perfectly on pit stops, and not have mechanical problems."

400

The number of laps Earnhardt ran at Bristol without power steering en route to victory in April of 1985.

Put any Winston Cup driver in a top car that's dialed in and they'll be fast. But only a few can continue to run at the front when the car's got problems. Nobody's better at manhandling and ill-handling race cars than Earnhardt, and there was no better evidence of that than this April afternoon at Bristol. "I've never driven a car which had lost the power steering and I hope I never have

EARNHARDT FANS

By Benny Phillips

They met at Bubba's Body & Paint Shop, early, as the racing rooster is crowing, and they fire up the Earnhardtmobile.

Usually they go with a 4-4-4-6 formation—four in the front seat, four in the back seat, four in the rumble seat and several six-packs tossed in for ballast.

The Earnhardtmobile is a 1972 Cadillac that Bubba bought for $500. He removed the trunk lid, welded in a rumble seat, and hack-sawed the roof off.

He blew out the rear window with a shotgun. Why? Because it was there.

The paint job is black and silver, of course, with silver flames and several No. 3 logos. Flying proudly from poles planted in the rear bumper are two black No. 3 flags.

If the boys get a good tailwind, they can make the drive down to Daytona from North Carolina in record time. They don't reveal their record time, but one among them claims they leave extra early to allow for being stopped by law-enforcement officers.

Charles is the chauffeur. He says he doesn't drink. He says a patrol officer pulled the Earnhardtmobile over one Saturday morning and advised the crowd, "Get this piece of [tin] off the road and I will not waste time writing you up."

Another cop stopped Charles and his buddies. "Man," the cop said, "you're sure lucky I'm an Earnhardt fan."

Earnhardt fans are a mystery. Nobody is real sure where some of them came from. Sociologists might speculate that either they migrated with the Vikings, or they're alien beings who were stranded when their UFO driver mistook a superspeedway for an intergalactic HoJo rest stop.

Some of these Earnhardt fans are—how can I say this?—less refined than the typical race fan. They're a lot like Earnhardt himself.

Average race fans, in Daytona for the 500 to escape the winter up North, are often seen batting beach balls around the infield. But before you just whip out a beach ball in front of those select Earnhardt fans, you might want to remind yourself of what Bubba did to that Caddy's rear window.

But, according to Charles, this group doesn't really get cranked up until the weather gets hot. According to a survey, the rowdy bunch "is not nearly as active until it gets hot, real hot, such as Daytona in July."

Charles says it's been a long time since the group had any trouble. In the past there were problems in the infield, although Lord knows it wasn't always "the group's" fault. Stuff happens.

Such as the time a 255-pound Bodine fan with an attitude problem sucker-punched Keith, one of the group. Bubba and Charles and his pals did some body work on the burly stranger.

OK, once Charles accidentally fishtailed the Earnhardtmobile in an Atlanta Raceway parking lot and splattered a white-suited attendant with mud, but that was an accident.

And there was the time at Rockingham when a security guard decided to join their party. He finally . . . er, well, fell asleep. Upon awakening, he found himself handcuffed to the Earnhardtmobile with his own cuffs.

Bubba says he is basically a peaceable guy. He says he keeps a chicken coop at his body shop with three hens and a rooster. The rooster, he says, had white feathers as a baby but incredibly, as it matured, its feathers turned black and silver.

Even the drive home is sometimes an adventure. Such as the time the Earnhardtmobile was pulled over by a female cop after a Talladega race. She put Charles, wearing his black leather walking shorts, through a 15-minute field-sobriety test.

She told him he passed. She also told him, "Nice shorts."

to again," Earnhardt said afterwards. "I had to pull on the steering wheel so much my right arm went to sleep."

Earnhardt's had so many similar performances, it's hard to keep track of them. Remember last year's Pepsi 400, when he was so loose he almost bought the wall on the last lap, yet still held off Ken Schrader and Sterling Marlin? And then there was the 1990 Southern 500, when he did hit the wall, with more than 200 laps to go. "The old gal slapped me when I got a little fresh with her," he said. "I felt like I was dragging a cow or tree hung up under the car."

He won anyway.

And even when he doesn't win, Earnhardt can squeeze a decent finish out of a lemon. "Sometimes they mean more," said Earnhardt, "because it means you work your butt off and you earn it."

"I was a little surprised Dale was able to come from the back to the front twice. I'm still shaking over that. The guy's awesome. . . . If the car is competitive, Dale can do the rest."

—Andy Petree, following the 1993 Busch Clash, his first race with the Childress team.

202

The number of points Rusty Wallace gained on Earnhardt in the span of two races—Dover and Martinsville—last fall.

At high noon on September 19, Earnhardt's advantage stood at 284 points. The fat lady was warming up in the wings. By sundown on the 23rd, Earnhardt's lead was down to 82 points and the fat lady had developed laryngitis. Finishes of 27th at Dover and 29th at Martinsville had done their damage.

Wallace, meanwhile, had finally come all the way back from his crash at Talladega. He'd strung together two wins and a second, and some of his best tracks were coming up.

At this point the Childress team showed their championship caliber. "We've had a couple of tough races," said Earnhardt. "Everybody has those, and we don't plan to have anymore."

Having seen the point race from every angle in the past, this team knew not to panic. Back in 1990, their position had been similar to Wallace's. In the summer they faced a triple-digit deficit to point leader Mark Martin. Then they won consecutive races at Darlington and Richmond, and suddenly the deficit was down to just 16 points. The tension grew through the fall, as Earnhardt and Martin finished within a spot of each other in four of five races—just as Earnhardt and Wallace did on three of four Sundays last October.

Both years, Phoenix was the turning point. In 1990, Earnhardt went into the desert trailing Martin by 45 points. He came out with a 6-point advantage after leading all but the first 50 miles. Martin, meanwhile, saw a possible top-five finish turn into a 10th when he had to pit late in the race after burning up his tires. Prior to that pit stop, he'd been desperately clinging to fourth.

That air of desperation seemed to stay with the Roush team in the season finale at Atlanta, when they chose to run a Robert Yates car rather than one of their own. It didn't help. They finished a lap down as Earnhardt cruised home third to win title number four.

In 1993, Earnhardt was the one clinging to a precarious lead going into Phoenix. And early on, it seemed he would suffer the same fate as Mark Martin in 1990. Wallace went to the front within 20 laps to collect his five bonus points, while Earnhardt struggled to stay in the top 10. Then, within 10 laps, everything changed. Wallace cut a tire and wore away his anti-roll bar as he scraped around the track. Six laps later, Earnhardt led two laps during a cycle of pit stops. He went on to finish fourth; Wallace faded to 19th. The fat lady got her voice back.

162.822 Miles Per Hour

Earnhardt's fastest lap during last summer's test session at Indianapolis.

After the Daytona 500, the inaugural Brickyard 400 in August will draw the most attention of any race on the 1994 Winston Cup schedule. So, naturally, Earnhardt is determined to win it, for

20 career wins at Daytona . . . *Mike Slade*

two big reasons: prestige and money. "I think [the purse] should be as much as the Indy 500," Earnhardt said at the speedway last summer. "That's between Bill France and Tony George. I had [George] on my boat on a little trip we took down to the Bahamas. I was working on him pretty hard and he said Bill had his hands tied. I said, " 'It's your pocketbook. Tell him to get his hands out of your pocket.' "

68

The number of lead changes in the 1984 Talladega 500. Earnhardt's first win with the Childress team.

At the time Earnhardt called it "the greatest race I've ever been involved in," but it offered just a hint of what was to come.

40

The longest losing streak of Earnhardt's career.

It began following his victory in Darlington's Rebel 500 on April 4, 1982, and lasted until the Nashville 420 on July 16, 1983. Ironically, that Darlington win had snapped a 39-race losing streak, stretching all the way back to October of 1980.

"When you've been accustomed to winning—winning big—and suddenly you're not anymore, it's going to make you feel down," said Earnhardt in Darlington's victory lane, his one oasis in an 80-race dry spell. "That's human nature. It gnaws at your insides. But my confidence wasn't shaken by losing."

30

The number of miles between Kannapolis and Charlotte.

It took Dale Earnhardt eight years, from his first short-track start in 1971 until his rookie Winston Cup season in 1979, to complete the trip

Oh, there were a few visits in the interim. In May of 1976, Earnhardt was at Charlotte Motor Speedway to run Walter Ballard's car in the World 600. Other than a few dozen short-track wins (and a legendary father who'd won a few hundred more), there was little to suggest that this rough-diamond racer would someday shine on NASCAR's superspeedways. He'd made just one Grand National (Winston Cup) start, a full year earlier, and had finished a forgettable 22nd.

Yet Humpy Wheeler, ever the visionary, saw something special in Earnhardt. "It has been a long time since I have seen a youngster so determined, so hungry," said Wheeler. "If nothing happens to sour his attitude, I think he's going to be a star in a few years—and a big one."

Wheeler had it wired. Just three years later Earnhardt was back at Charlotte, this time with an established team and one Winston Cup win already in his pocket. He led the 600 on 11 occasions and finished third before an estimated 135,000 fans. "I was absolutely awed by the crowd when I was introduced before the race," said Earnhardt. "I think all of my hometown was here. We didn't win, but I think they know we were here."

Winner Darrell Waltrip sure did. "It looks like when everybody else is gone, Earnhardt is one I'll be fighting with down the road," said Waltrip.

Richard Petty, who barely held off Earnhardt for second, knew it too. Said Petty, "Where you been, boy?"

Fourteen years later, Earnhardt would create a stir at Daytona by echoing this line in his description of Jeff Gordon. Maybe he remembered May 27, 1979—the day "that Earnhardt boy" arrived.

"I think Ralph really believed someday Dale would do something like he did today," Dale's mother, Martha, said after the 600. "In fact, a lot of times I wonder that Ralph might well know what's going on."

"We got our heads together and Richard ended up getting out of the car, painting it blue and yellow, and letting me drive it. At the end of [1981] our agreement was over and I went with Bud Moore, but I was never really satisfied with the performance of the Ford engines at the time . . . All the time Richard Childress and I had been talking back and forth. I felt Richard, myself, a good sponsor, and Chevrolet could bring along a championship."

27

The number of Backfire letters SCR *printed in 1987 following Earnhardt's "Pass in the Grass" incident during The Winston.*

Fifteen were negative, 12 positive. Best line: "Get him off the track and into a counseling program."

23

The number of cars involved in a first-lap crash in the 1990 Pepsi 400.

Earnhardt's was not among them. He'd started on the outside front row, shot into the lead, and was out of harm's way when Derrike Cope, Richard Petty, and Greg Sacks came together while running three-wide behind him, blocking the track. Earnhardt went on to an easy win.

In February of that same year, there'd been another 23-car crash at Daytona, this one in the Goody's 300. Earnhardt won that race, too.

"The whole world wrecks, and he drives right through without a scratch," Rusty Wallace once told *SCR*. "Dale Earnhardt is definitely the guy who gets through the most. As far as knowing when to go, where to go, and when not to go, it's him."

And Wallace said that *before* the two Daytona wrecks.

20

The number of races Earnhardt has won at Daytona.

If you throw out the 500 (and many times he's wanted to), Earnhardt has had a brilliant record at Daytona. In fact, he's won more races there (in the Busch Series, the Busch Clash, IROC, 125-mile qualifiers, and the Pepsi 400) than anyone else. His first win came in the 1980 Busch Clash. It was also

. . . But none in the 500. *Nigel Kinrade*

his first superspeedway win in a Winston Cup car. To get it, he had to slingshot around Darrell Waltrip on the last lap—not a maneuver for the faint of heart. But he pulled it off. Wrote *SCR* editor Dick Berggren, "Earnhardt has let it be known that he does not want to be referred to as a rookie any longer."

18

Earnhardt's age when he was married for the first time.

Earnhardt has been married three times, but the first two were short-lived. He's been married to Teresa since November of 1982, and he's frequently credited the longevity of their marriage with helping to stabilize his career. "I would rather have been married one time and lived happy ever after, but it didn't happen that way." Earnhardt told *SCR*'s Father Dale Grubba in 1985. "I was fortunate to have found Teresa."

Earnhardt has four children.

16

The number of starts it took Earnhardt to win his first Winston Cup race. No kidding: it happened at Bristol on April Fool's Day, 1979.

Once he got in a competitive car, it didn't take Earnhardt long to show he belonged in the major leagues. And after that first win, he remembered his father, the man who had toiled so long in the minors. "I know that somewhere there's a fellow that's got a big smile on his face and is mighty proud and even happier than I am, if that's possible," said Earnhardt. "I am determined to keep him smiling for a long time."

"I can take a race car and carry it, and Rusty can, too...It's always been like that, ever since he got into a competitive car. We've always enjoyed going after each other, and I think it's always going to be that way."

13

The number of laps Earnhardt completed at Charlotte in October of 1989 before the camshaft broke, leading to a last-place finish.

Should Earnhardt fail to equal—or surpass—Richard Petty's seven championships, he'll remember 1989 as the one that got away. In fact, the 1989 season marked the only time he's finished as the series' runner-up. Every other time he's been a contender for the Cup, he's grabbed it. He led Rusty Wallace by 165 points in August of 1989, but a series of autumn disasters left him playing catch-up heading into November. His last-lap scuffle with Ricky Rudd at North Wilkesboro—which saw a possible win or a sure second turn into a 10th at the touch of some sheet metal—and his run-in with Rusty at Rockingham got the most publicity, but that broken cam at Charlotte was the killer. "You can't do anything about metal breaking," said Earnhardt. "If it gives up, it gives up."

Which was something the team never did. Said Richard Childress, "We don't know the meaning of the word quit."

12

The number of points Earnhardt ultimately came up short that year.

But he had no regrets. "The day before [Rockingham] we were talking about the Wilkesboro-Rudd accident, you know, racing side by side, banzai down into the corner," Rusty Wallace told *SCR* after taking the title. "I asked Dale, 'Would you do it all over again, would you run him hard [or] would you back off?' He said, 'Hell, no, I'm a race car driver. I'm going to win. I'd race him. If you're side by side, both guys going into a corner, it's a real close move and you keep going. Hell yeah, I'll keep going.' "

Still, you have to believe the Rudd wreck stayed with Earnhardt. After running second to Wallace at Rockingham last February, he remarked, "I'd be the bridesmaid all season if we could win the championship."

And while that's not quite the way it worked out—Earnhardt did win six times en route to the title—Wallace was the season's biggest winner, with 10 victories.

11

The number of times Earnhardt and Wallace have finished first and second (or vice versa) on the race track.

Earnhardt has had no shortage of rivals through the years—from Cale Yarborough to "that Gordon boy"—but there's been something special about his relationship with Wallace. It started at Atlanta in 1980, when Wallace made his Winston Cup debut. Driving a Chevrolet for Roger Penske, Wallace finished a brilliant second—to Earnhardt. It's as if they were destined to shadow each other. "The determination factor with Rusty and I is really up there," Earnhardt acknowledged last fall. "It's probably unmeasurable. To have him racing close to me or me racing close to him, it makes us that much more determined.

"You've got to have somebody good to race to have a real good time at it," Earnhardt continued. "Darrell and I raced through several years and Bill and I, and now Rusty comes along. It's become a mutual-respect type of race when you go in and race Rusty. We race each other to the limit."

10

The approximate number of times Wallace turned over at Talladega after contact from Earnhardt last May.

This is the flip side of their rivalry. Yes, they're friends and yes, there's a mutual respect. But, as Earnhardt said, they race each other to the limit—and sometimes beyond. Their battles have occasionally resulted in battered race cars, bruised egos, and angry words. Said Wallace in 1989, after the lapped car of Earnhardt ran him out of the groove at Martinsville and allowed Darrell Waltrip to sneak past for the win: "Two can play at that game, if he wants to." Said Earnhardt last September, after Wallace rear-ended slowpoke Hut Stricklin on a restart, touching off a crash that collected Earnhardt, among others: "I don't know what happened, but I know the 27 didn't miss a gear."

So do these two men in black show their true colors at such moments? Is their friendship just a PR put-on?

No. There was no mistaking the concern on Earnhardt's face when he stopped in the Talladega tri-oval to check on Wallace's condition following Rusty's harrowing wreck. "He's the tough, macho-type image," Wallace once told *SCR*. "He realizes that's his image, and he likes to play it. But he's just a big old softy. He'd do anything for you."

The most Winston Cup races Earnhardt has won in a season: 11, in 1987. *Mike Slade*

"We'll put pressure on to go for Petty's record and tie it, but we've got that sixth one behind us now . . . That's what it's all about."

Number of times Earnhardt and Wallace have finished first and second (or vice versa): 11. *Harold Hinson*

That soft side of Earnhardt was never more visible than at Talladega. "I wouldn't have done that to anybody on purpose, especially him," said Earnhardt. "Rusty and I are too good of friends for that."

9

The number of car numbers Earnhardt has used in his Winston Cup career.

He's become so closely associated with that trademark No. 3 (and incidentally, that menacing white-on-black number is trademarked), it's hard to imagine Earnhardt ever running anything else. But before he hooked up with Richard Childress in 1984, Earnhardt used the numbers 2, 8, 15, 19, 30, 77, 96, and 98 at various stages of his Winston Cup career.

8

The number of grades Earnhardt completed in school.

"I was young and dumb and should have listened to my father," Earnhardt once said of his decision to drop out. "You work hard and things come to you, but still there are days through my championships and my wins that I wished I had a better education and made better decisions."

7

The number of the late 1992 Winston Cup champion, Alan Kulwicki.

Emotionally, 1993 was the most difficult of Earnhardt's six championship seasons. "We had a tough year at times," Earnhardt said after clinching the title at Atlanta. "We had a great year at times. We had a real sad year in losing our champion, Alan Kulwicki, and our good friend Davey Allison. It's a tough year to review and think about, but it's great to end it on a super note winning this championship. Putting that behind us, we'll have great memories of Alan and Davey as we honored them today making the victory lap."

That moment—Earnhardt and Wallace circling Atlanta in tandem, flying the flags of Alan Kulwicki and Davey Allison—served as an eloquent summary of the 1993 Winston Cup season.

6

The number of manufacturers Earnhardt has driven for.

He's been a Chevy man exclusively since 1984 (not counting non–Winston Cup rides, such as IROC), but before that Earnhardt sat in just about every type of Cup-legal equipment available. He ran Fords for two full seasons under Bud Moore, but soured on the association when he recorded far more DNF's (18 in 1982 alone) than wins. He ran Pontiacs in 1981, a combination of Chevys and Oldsmobiles in 1980, and combination of Chevys and Buicks in 1979. And—Mopar fans take note—way back on May 25, 1975, when Earnhardt made his Winston Cup debut in the World 600, he was driving a 1974 Dodge owned by Ed Negre.

5

The number of championships Earnhardt has won with Richard Childress.

Sure, they're a juggernaut now. But when the easy-going owner first teamed up with the brash young driver, a lot of questions awaited answers. As *SCR* editor Dick Berggren wrote in his 1984 preseason analysis: "Dale Earnhardt remains one of my favorite drivers to watch. He's ever so aggressive, runs the wall as tight as Petty (nobody runs any closer to disaster than Richard and Dale), is a brilliant strategist and gets everything there is out of a race car. However, I see Richard Childress, Earnhardt's new car owner, as a different personality than Earnhardt. If they function well together, hang on. Earnhardt will be a solid contender. Childress, on the advice of Junior Johnson, has been building his own equipment for several years and, as Ricky Rudd proved in 1983 with five poles and two wins, Childress cars go. This team will either fall flat on its face or it'll be right in there. There is no middle ground."

Championships with Kirk Shelmerdine: 4. *Staff photo*

4

The number of championships Earnhardt won with Kirk Shelmerdine as crew chief.

When Shelmerdine, who'd been with Childress even longer than Earnhardt, announced he was leaving at the end of 1992, some saw that as the beginning of the end for the tight-knit team that had won back-to-back titles in 1986–1987 and again in 1990–1991. But, Shelmerdine insisted, "We're not doing anything more wrong now than we were doing right five years ago when we were winning everything." He said he was leaving to pursue his own racing career, not due to dissension.

Subsequent events bore this out; Shelmerdine was among those Earnhardt thanked from the stage at the Waldorf last December. And he plans to run the Busch Series in 1994.

3

The last number a racer wants to see in his rearview mirror when the white flag flies.

Earnhardt's reputation as an intimidator tends to put other drivers in a defensive posture, subconsciously or otherwise. As Richard Petty once put it,

"I feel good. I'm racing good. I'm driving good. . . . If I just keep the pace up, I think I'll be OK." *Mike Slade*

"When Dale Earnhardt is running beside you, you will react a little differently than if Bill Elliott or Terry Labonte has just pulled up alongside of you."

2

The number of penalties Earnhardt overcame to win the 1993 Coca-Cola 600.

Because it was so difficult, this was easily Earnhardt's best win of the season. First, NASCAR slapped him with a 15-second citation for speeding on pit road. A mere annoyance. There was plenty of time to get back in the race, and Earnhardt did. Later, however, a yellow flag flew during a round of green flag pit stops and left Earnhardt at the tail end of the lead lap. He would need a caution flag to help him catch up, and he got one when Greg Sacks spun less than 100 laps from the finish. Unfortunately, NASCAR decided that Sacks spun because Earnhardt ran into him—a charge Earnhardt denied. "No, I did not hit him," said Earnhardt. "I may have got a-gin' him . . ."

The semantics got even better: "After I didn't hit him," said Earnhardt, "I had no idea they would penalize me."

Earnhardt immediately got his lap back on the restart. When another yellow again bunched the field, Earnhardt had his opportunity. He bulled past Ernie Irvan and went on to win. "It's really satisfying to win after getting penalized," he said. "As competitive as everybody is, you think you might not be able to make it up. We made it up fairly easy."

1

The number of races Earnhardt won in 1992.

Having taken three of the four Winston Cups handed out in the 1990s, Earnhardt is clearly the driver of the decade so far. And that makes it easy to forget just how awful his 1992 season was—at least by his standards. His 15 top-10s was the fewest he'd had since 1983; six top-fives was his fewest ever. His 12th-place finish in the standings was his worst since 1982 and marked the biggest point plunge by a defending champion since 1971. Other teams might have splintered after falling so far so fast, but not this one. "Our guys bounce back better," Earnhardt insisted during the midst of that dismal 1992 season. And their remarkable turnaround in 1993 certainly proved him right.

0

The number of times Earnhardt has won the Daytona 500.

OK, so that's a little obvious. But the subject will dog Earnhardt throughout Speedweeks 1994, just as it did in 1993, 1992, 1991, 1990. . . .

Nineteen-ninety was the year that really put Earnhardt under the microscope. Darrell Waltrip had won the year before, ending an 0-for-16 Daytona 500 drought and leaving Earnhardt next in line for the can't-win-the-big-one blather. Earnhardt responded with one of the most dominant performances of recent years, only to hit a piece of debris less than a mile from the finish line and . . . well, you don't need Paul Harvey to tell you the rest of the story.

You'll hear it again and again. So will Earnhardt. And until 12:15 P.M. on February 20, he'll sit and smile and say about the only thing he can say. "It's a new year. It's a new deal. Win or lose, Daytona is going to come next year and it's going to start all over again, no matter what happens this year."

8

SIX AND COUNTING!

Dale Earnhardt Has Won Six Winston Cup Championships, But Will He Ever Win Daytona?

BY BOB MYERS
Circle Track March 1994

Irrepressible Dale Earnhardt and the Richard Childress Racing juggernaut win the Winston Cup championship. They sip the heady champagne and collect the spoils of incomparable success. Then they go to Daytona International Speedway and win most of the preliminaries. Earnhardt is hyped as the Daytona 500 favorite, and then he loses it, just like two of the past four races, on the very last lap. That's the scenario that has been played out in recent years. Will it play again on February 20?

Earnhardt has become heir to Richard Petty's throne in the toughest, richest, and most popular stock car racing series. With six championships (five with the Childress wonders), 59 victories, and about $20 million in NASCAR earnings, he won practically everything. En route to the $1.5 million championship and a gross of about $3 million in 1993, Earnhardt has won six points races, the Winston All-Star race, the Busch Clash, and a 125-miler at Daytona.

But never the Daytona 500. It is frustrating to know, especially after another bountiful year, that triumph in the sport's premier race has eluded Earnhardt and his team. Upsets from stunning to mild by Derrike Cope, Ernie Irvan, and Dale Jarrett, yearlings in the sport compared with Earnhardt's 15 full seasons, have come in three of the past four Daytona 500s. In 1990, Earnhardt lost due to a cut tire on the last lap. Cope

Six and counting; can Dale pass Petty for the all-time championship season total? *Nigel Kinrade*

Quick pit stops were a key element for Earnhardt's drive to his sixth Winston Cup Crown. *Nigel Kinrade*

capitalized on Earnhardt's misfortune and won. In 1992, Jarrett nipped him in a last-lap showdown. For the 16th time, Earnhardt faces his nemesis in his quest for the missing jewel.

"They ought to make it the [Daytona] 499," says Earnhardt, grinning menacingly under his gunslinger's mustache.

"The world won't end if I never win, but I'd give up all the preliminaries if I could."

The team is probably the best prepared it has ever been for the 500 and the new season, says crew chief Andy Petree, whose first big-league title came in his first year as the team's leader. "Our new Daytona car tested great [at Talladega] under a few rule changes in early November," he says.

The rule changes for Daytona include a roof height of 51 inches, a 1-inch increase in the left side quarter panel height from 33 inches to 34 inches, a 2-inch increase on the right side from 33 inches to 35 inches, and an increase in tread width from 59.5 inches to 60.25 inches, along with the mandatory carburetor restrictor plate.

"The changes will probably slow the cars another two- to three-tenths," says Petree.

It's not preparation and motivation that the team needs at Daytona, it's luck.

"After the tests, I came back to the shops and told [crewman] Will Lind how well the car performed," says Petree. " 'But did it look lucky?' he asked."

Earnhardt admits there is self-induced pressure to win not only the Daytona 500, but to equal Petty's record seven championships. "Richard Petty will always be the king," says Earnhardt. "I can't touch his 200 wins and it doesn't look like [I can touch] his seven wins in the Daytona 500, but now I've got a good shot at tying and breaking the record for championships. Winning the sixth title was no different from the others except it puts me in a position to tie the record.

"Whether you're racing or playing ball, you always look up to somebody. That's the way I am with Richard. He has done it all and that's what racing is all about."

When Earnhardt went after Cale Yarborough's record three consecutive titles in 1992, he had just finished the worst of 10 full seasons with Childress. But five out of the six times he has finished in the top five in the Daytona 500, he has gone on to win the championship.

The champion carried the load for Chevrolet in 1993 and, on paper, their burden will be great in 1994. Although Chevrolet's 10th Manufacturer's Championship in 11 years was decided by Ricky Rudd's second place in the season-ending Hooters 500 at Atlanta, Earnhardt contributed 100 of Team Chevy's 191 points. Archrival Ford had 190 points and Pontiac had 189, making it the closest manufacturer's competition in history.

Chevrolet won nine of the first 18 races, including Earnhardt's six, but was blanked in the final 12, finishing last to Pontiac's 11 and Ford's 10. Rusty Wallace had a glorious year in Penske Racing Pontiacs. Even though his courageous comeback from two spectacular wrecks during the season had him 341 points down with 12 races left, he fell 80 points short in the classy championship showdown.

With 10 victories, three straight in the spring, and five in the last eight races, including the finale at Atlanta, Wallace won a third of the season's races and almost a third of his career 31. Kyle Petty won the other race, making this Pontiac's best season since 1962, perhaps one to savor.

Mark Martin and Ernie Irvan saved Ford's season. Ford had three wins when Martin, who finished with five, began a four-race streak in Roush Racing Thunderbirds. Irvan, after defecting from Morgan-McClure Chevrolets to Robert Yates Racing 'Birds, added two.

Still, Chevrolet wound up with most of the spoils. Ken Schrader, although winless, added the Busch Pole Award with six, and Hendrick Motorsports stablemate and first-year sensation Jeff Gordon swept Rookie of the Year honors for the Bow Tie brigade.

Wallace, Earnhardt, Martin, and Irvan (who won three) won 24 races altogether, leaving one each for the six other victors, including Ricky Rudd, Morgan Shepherd, Geoff Bodine, and the late Davey Allison.

If 1993 means anything, it's that Ford looks awesome in 1994, at least on paper. The switch by Wallace/Penske and Rudd, as an owner/driver, to Ford's ranks, along with Irvan's presence, represents 12 General Motors wins in 1993, leaving only two 1993 winners in Chevrolets—Earnhardt and Daytona 500 champion Dale Jarrett. Ford lost Sterling Marlin to Chevrolet and Bill Davis Racing, and driver Bobby Labonte to Pontiac during the shuffle. Fords proliferate, with at least 20 drivers and teams fielding 'Birds, including the nucleus of Martin, Bill Elliott, Geoff and Brett Bodine, and Shepherd.

"We were disappointed overall," says Lee Morse, Ford's assistant director of Motorsports, "but considering our slow start and the devastating loss of Alan Kulwicki and Davey Allison, I think our comeback was spectacular. With the new additions, the outlook is bright in 1994."

Chevrolet counters with Earnhardt, Harry Gant, Jarrett, Waltrip, and the Hendrick triumvirate of Schrader, newcomer Terry Labonte, and Gordon. Kyle Petty and Michael Waltrip lead what was, at the end of the season, a five-team Pontiac contingent.

The bottom line, though, is that Chevrolet has Earnhardt. The team rebounded from a terrible year, one win and 12th ranking in points in 1992, to win it all in 1993. "It's unbelievable how this team keeps on ticking," Earnhardt says. "Andy came on board and fit in like a glove, bringing fresh ideas and motivation. He said when we interviewed him that we could win championships together. That impressed us. We did and we took that rookie to New York for the awards banquet

ANATOMY OF THE 1993 WINSTON CUP CHAMPIONSHIP

	Earnhardt's Position*			Wallace's Position			
Race	**St.**	**Fin.**	**Points Position**	**St.**	**Fin.**	**Points Position**	**Earnhardt's Margin**
Daytona 500	4	2	1 (tie)	34	32	32	+148
Goodwrench 500	7	2	1	10	1	8	+103
Pontiac 400	11	10	2	13	2	5	+67
Motorcraft 500	2	11	1	1	3	3	+27
TranSouth 500	1	1	1	3	5	2	+57
Food City 500	6	2	1	1	1	2	+47
First Union 400	21	16	2	9	1	1	–18
Hanes 500	21	22	2	5	1	1	–101
Winston 500	1	4	2	24	6	1	–86
Save Mart 300	1	6	1	6	38	2	+20
Coca-Cola 600	14	1	1	8	29	2	+129
Budweiser 500	8	1	1	4	21	2	+209
Champion 500	5	11	1	10	39	5	+298
Miller 400	6	14	1	15	5	3	+269
Pepsi 400	5	1	1	17	18	3	+345
Slick 50 300	24	26	1	33	1	3	+250
Miller 500	11	1	1	18	2	3	+260
DieHard 500	11	1	1	32	17	3	+333
Bud at The Glen	5	18	1	6	19	3	+341
Champion 400	5	11	1	10	6	3	+324
Bud 500	19	3	1	2	2	2	+309
Southern 500	6	4	1	11	3	2	+304
Miller 400	8	3	1	3	1	2	+284
SplitFire 500	9	27	1	1	1	2	+180
Goody's 500	7	29	1	4	2	2	+82
Holly Farms 400	10	2	1	11	1	2	+72
Mello Yello 500	9	3	1	21	4	2	+82
AC-Delco 500	22	2	1	18	1	2	+72
Slick 50 500	11	4	1	6	19	2	+126
Hooters 500	19	10	1	20	1	2	+80

*Earnhardt also won the Busch Clash at Daytona and The Winston All-Star Race at Charlotte and taught him how to party. Every man dug deeper and that's what turned a bad year into another title. I'm really proud of my team."

The feeling is certainly mutual. "I made the change because I wanted to be on a championship team," says Petree, who replaced 46-race-winner Kirk Shelmerdine. Petree's previous high finish in points was fourth in 1991 and 1992 as Gant's and the Leo Jackson team's leader. "After a year with the team, I think the key to its success is that it takes on Earnhardt's personality.

"He stays motivated all the time, like a wide-eyed rookie who has never won a championship. He never says die. The team admires him for staying motivated even though he has won so much. He's just as hungry now to win the championship next year as he was to win the first. That rubs off on the team. That's in addition to how good Earnhardt is and I don't have to tell anybody that.

"We want a Daytona 500 [win] as much as he does. Losing last year's was probably the biggest disappointment of my career. That was my first race with the team and I told the guys I really felt like one of the family. The team has dominated and come so close so many times and that compounds the disappointment. It also makes us more determined to win one for Earnhardt. It's hard to fathom that he hasn't won it. We're all focused on changing that this year."

Team owner Richard Childress says the poor 1992 season was a sacrifice to regroup and that 1993 was the result of that effort. "We got behind in 1991," says Childress, "fighting for [and winning] the championship while everybody else was preparing for the new season. We ran the new cylinder heads all season, although they weren't mandatory until August, and we experimented with some other engine stuff that didn't work. All of that experimenting paid off in 1993. Andy came in and it seemed he'd been with us 10 years. He established a wonderful working relationship with Dale and the crew. There was almost no learning curve. Some changes in responsibility [duties of Terry Eldridge, Bobby Hutchens, and Cecil Gordon, for instance] in the shop helped. The biggest thing is that Dale and all 36 others contributed 100 percent."

The team—among its championship-proven crewmen Lind, David Smith, Chocolate Myers, engine chief Eddie Lanier, and engine specialist Danny Lawrence—had only one DNF, a broken rear-end gear at Martinsville, and no complete engine failure en route to its fifth title in eight years. Earnhardt led the points competition in all but four races, regaining the lead for keeps when Wallace finished 38th at Sears Point, the 11th race. Jarrett was the only other points leader.

Wallace was 148 points in arrears after the first race, flipping wildly in the Daytona 500. However, he built a 101-point lead during a three-race winning streak in the spring and had an 86-point lead after he was tapped by Earnhardt and flipped again, breaking his wrist in the Winston 500 at Talladega.

Wallace didn't accuse his close racing buddy, but Earnhardt took the blame. At Dover, Earnhardt mildly scolded Wallace for triggering a wreck that damaged his Chevrolet and, coupled with Wallace's victory, cost 104 points. For the most part, the battle between the friendly rivals was a class act, devoid of roughhousing and mind games.

Relentless, Wallace, with two races left, and with four wins in six races and eight straight finishes ahead of the leader, chopped the margin to a low of 72 points with his victory at the AC-Delco 500 at Rockingham. He lost 54 points to a flat tire and suspension damage at Phoenix and regained 46 for the final deficit of 80 with his romp and Earnhardt's 10th at Atlanta. Martin finished third in points, 376 behind, Jarrett fourth (-576), and Kyle Petty fifth (-666).

"It was frustrating to do what we did and not get closer than 72 points, and it is unbelievable that we won 10 races and didn't win the title," says Wallace. "But we had a wonderful year with a great team and that's not soured by losing the championship. I've said many times that Earnhardt is the best driver on the circuit, so we know we got beat by the best. I got myself mentally prepared to lose the title and I'm mentally prepared to win it next year. I told the media before last season started that we we going to win the championship. Then Buddy Parrott, our crew chief, told us we were going to win 10 races. We sort of snickered. One out of two isn't bad. I can't see anything but better things happening to us in 1994, unlike 1990, the year after my championship season, when the Blue Max team fell apart."

Crusty ol' Earnhardt was misty-eyed and his words dripped with humility as he talked about how miraculous it was for a boy from the small textile town of Kannapolis, North Carolina, who never dreamed of racing Winston Cup, to sit atop the world's premier stock car series. He also spoke of how incredible the team was to keep on, and how bittersweet the year was with the deaths of Kulwicki and Allison.

Tribute was paid to the fallen comrades before and after the Hooters 500. At the end, Earnhardt and Wallace, their war over, drove counterclockwise laps carrying flags bearing the late drivers' car numbers, endearing themselves to the estimated crowd of 120,000. "We'll put our championship and Rusty's gutsy comeback effort behind us and have great memories of Alan and Davey," says Earnhardt.

Perhaps those who believe in numerology notice, as did waitresses Cindy Ritch and Shannon Payne at Nick's Tasty Platter Restaurant in Charlotte, where racing is on the menu daily, a rather bizarre combination of numbers in the finish of the Hooters 500. Winner Wallace drives car No.2, runner-up Rudd No.5. The sum is 7, Kulwicki's number. Darrell Waltrip, No.17, was third and Elliott, No.11, was fourth. The sum is 28, Allison's number. Subtract 2 from 5 and the remainder is 3, the champion's number.

Earnhardt and crew celebrate a victory at Darlington. It was win #1 in the 1994 Winston Cup title chase. *Nigel Kinrade*

9

SEVEN FOR EARNHARDT

1994 Winston Cup Champion

BY BOB MYERS
Circle Track March 1995

Newly crowned for the record-tying seventh time, Dale Earnhardt is in the enviable position of setting the record for Winston Cup championships, and figures he has five or six years to accomplish the mission with the dynamic Richard Childress Racing team.

"If I stay healthy and everything goes according to plan, I'm going to race until [the year] 2000," says Earnhardt, 43. "Contrary to rumors that I'm going to organize a team in 1996, I plan to drive for Richard Childress until I retire.

"We've won six titles together. He and this team are a part of me in business and friendship. They enabled me to win seven championships and to equal Richard Petty's record. That's a tremendous accomplishment for us. When I started, I never expected to see one. But we don't want to settle for a tie. I think we'll be more determined than ever, and I think there's time to get the eighth."

Childress heralds Earnhardt's commitment to the famed Black Car. "We've had a great 11 full seasons together," says Childress. "He continues to prove that there's nobody better. He's getting smarter and better with age, and that's what we saw in 1994.

"Preparation is the key to winning championships. We've thrived on that. We've never been better prepared than [we were] last year. We began preparing for next season before the race was over and after we clinched it, we started busting our rear ends. Two straight championships drain a team physically and restrict preparation for the next year, so we're trying to guard against that. We're going after the eighth title, but not because we can break Petty's record. We go after all of them."

Dale Earnhardt matched Richard Petty's record seven Winston Cup championships with his 1994 title. *Nigel Kinrade*

The famed Black Car posted four victories and 25 top-10 finishes during the 1994 season. *Nigel Kinrade*

And these contemporary masters have reached the ultimate in Winston Cup in four out of the past five years and six of the past nine (1986, 1987, 1990, 1991, 1993, and 1994) after Earnhardt started this string of jewels in 1980 with defunct Rod Osterlund Racing. The titles have come in pairs three times, but never three in a row. The goal again in 1995 is a "three-peat," accomplished only once by retired Cale Yarborough in Junior Johnson's cars, 1976 to 1978.

There is the uncertainty of changing Chevrolet racers from the trusty Lumina to the sleek new Monte Carlo—a reincarnation of the winningest nameplate in NASCAR's big league. The Lumina was introduced during the 1989 season, when Earnhardt lost the title to Rusty Wallace and Pontiac by a mere 12 points, the second closest finish in Winston Cup. However, Earnhardt has won four championships and 28 of his career victories in 170 starts in the Lumina. He needed 311 starts to win his first 35.

"I think the Monte Carlo is going to be very competitive," says Childress, "but it's not going to have an advantage on Ford."

In early tests, GM technicians and teams concentrated on reducing the Monte Carlo's aerodynamic drag and to offset a high decklid and rounded rear end. There's an imbalance of downforce, less on the front, more on the rear. "The challenge for us is to get the car's drag down," says GM motorsports spokesman Don Taylor. "One problem is that the raised decklid puts more spoiler into the air.

"The Monte Carlo probably lends itself more to the intermediate and short tracks. We're hoping it will be at least equal to the Lumina on speedways. There's a lot of pressure to live up to expectations based on the history of the old car. There's been so much public attention [that] people are expecting it to be a winner right out of the box."

"NASCAR talked about making all the cars the same length front and rear," says GM engineer Terry Laise, "but there wasn't time to get that worked out for 1995."

"Cars don't win championships," interjects Lee Morse, Ford Motor Company's assistant motorsports director. "They're won by drivers, teams, and technical and financial assistance."

Earnhardt says the prospect of christening a new car with two records, an eighth title, and three straight championships is mind-boggling, but no more so than what a boy with lots more desire than means from the small town of Kannapolis, North Carolina, has accomplished already.

That's reminiscent of another boy from the even smaller North Carolina community of Level Cross who became king; in fact, he still is and will always be king in Earnhardt's herd. "Richard Petty is the king," says Earnhardt. "He's done it all. He's done a lot more than win 200 races. He pioneered the sport and got us where we are. He is among numerous people who helped me learn how to race and to get where I am. I just tried to put tying his record out of my mind, because I knew how much it would mean to me. Now, I'm just proud and honored to be in the same league with him in terms of the record."

Earnhardt is and isn't in the same league with Petty, whose titles came in 1964, 1967, 1971, 1972, 1974, 1975, and 1979, 13 years before he retired as a driver in 1992. Petty won 93 of 277 starts (33.5 percent) in his championship years; Earnhardt won 44 of 208 starts (21 percent), a career-high 11 in 1987.

Petty's 27 victories, including 10 straight in 1967 and 21 in 1971, and his 35-year career 200 seems untouchable under the present format. However, Earnhardt, whose championship era

CHAMPIONSHIP YEARS

Dale Earnhardt

Year	**Races**	**Wins**	**2nd**	**3rd**	**4th**	**5th**	**6-10**	**Earnings**
1980	31	5	3	4	3	4	4	$ 588,926
1986	29	5	5	3	1	2	7	1,168,100
1987	29	11	5	1	2	2	3	2,069,245
1990	29	9	3	3	1	2	5	3,083,056
1991	29	4	3	4	1	2	7	2,396,685
1993	30	6	5	3	3	0	4	3,353,789
1994	31	4	7	8	1	2	5	2,778,895*

*Includes only season earnings and $1,250,000 R. J. Reynolds bonus.

Richard Petty

Year	**Races**	**Wins**	**2nd**	**3rd**	**4th**	**5th**	**6-10**	**Earnings**
1964	61	9	14	11	0	2	5	$ 98,810
1967	48	27	7	2	1	1	1	130,275
1971	46	21	8	7	2	0	3	309,225
1972	31	8	9	5	2	1	2	227,015
1974	30	10	8	4	0	0	1	299,175
1975	30	13	5	3	0	0	3	378,865
1979	31	5	7	2	4	5	4	531,292

began as Petty's ended, has grossed more than $15 million compared to Petty's $2 million in championship-year earnings. Petty's largest championship bonus was $150,000 in 1979, when he grossed $531,292. Earnhardt's minimum bonus for 1994 was $1.25 million, and his earnings are expected to exceed $3 million for the third time when bonuses are counted.

Petty's reaction to having his record matched is positive. "It doesn't make a whole lot of difference to me," he says. "I never thought much about it. I think it's good, especially the national exposure for the sport. Dale is our leader and the leader needs to be the champion.

"The eras are entirely different. They don't really compare. It doesn't diminish what I did. People run for the championship now because of the big money, prestige, and exposure. They didn't [run for that] in the '60s and '70s. They concentrated on winning races. I could have won three or four more titles if I had focused on them. The only time we got real concerned about the championship back then was if there were a few races left and we had a shot."

The Richard Childress Racing pit crew was superb as always in 1994. *Nigel Kinrade*

Earnhardt and Childress certainly focused on the championship in every race, and Petty's record was an added incentive in 1994. "We approach every race with the title in mind," says Earnhardt. "We don't stroke by any means, but it's essential to finish practically every race to win the championship.

"The end of the tire war, with Hoosier's withdrawal, is a real plus for our team and the sport going into this season. In every race last year, nobody knew what to expect. This will put all the cars on the same track. The outcome of races will be determined by the car and driver, not by the brand of tire they're running. I've won seven championships with Goodyear tires on my car and hope to win more on that brand. We just don't need a tire war."

Number 3's win in the AC Delco 500 at Rockingham clinched the championship. *Nigel Kinrade*

Earnhardt's tactics were on the money. More impressive than his four wins was his amazing consistency, reflected by an average finish of 7.1379 (with one DNF). After 29 of the 31 races, Dale clinched the championship at Rockingham with an insurmountable 448-point lead over Rusty Wallace. Wallace's average finish then was third-lowest at 12.0345.

Earnhardt received a pivotal, perhaps decisive break, when top challenger and Ford rival Ernie Irvan, who won three races and a series-tying five poles in 20 starts, wrecked in practice on the eve of the August 21st race at Michigan and was sidelined indefinitely. Irvan had led the points battle in Yates Racing's Thunderbird on four occasions for 13 races and trailed by only 27 points at the time of his accident. Oddly, the Michigan race marked Earnhardt's only DNF before he clinched, and neither Wallace nor Martin got closer than 203 points.

"We had a great battle going with Ernie, and it might have gone to the wire," Earnhardt says. "After that, all we had to do was keep Rusty at a distance and make it to the end."

ANATOMY OF THE 1994 WINSTON CUP CHAMPIONSHIP

Race	Earnhardt Fin.	Pts.	Position in Points	Wallace Fin.	Pts.	Position in Points	Earnhardt's Margin
Daytona 500	7	151	6	41	40	41	+111
Goodwrench 500	7	151	3	Win	185	13	+77
Pontiac 400	4	165	2	2	175	8	+67
Purolator 500	12	127	3	24	91	11	+103
TranSouth 400	Win	185	2	33	69	13	+219
Food City 500	Win	185	1	7	151	10	+253
First Union 400	5	155	1	2	175	6	+233
Hanes 500	11	130	2	Win	185	4	+178
Winston 500	Win	180	2	33	64	7	+294
Save Mart 300	3	170	2	5	155	6	+309
Coca-Cola 600	9	138	2	2	180	3	+267
Budweiser 500	28	79	2	Win	180	3	+166
UAW-GM 500	2	175	2	Win	185	3	+156
Miller 400	2	175	2	Win	185	3	+146
Pepsi 400	3	165	2	26	85	3	+231
Slick 50 300	2	175	1	3	165	3	+241
Miller 500	7	146	1	9	134	3	+244
DieHard 500	34	66	2	42	37	4	+273
Brickyard 400	5	160	1	4	165	3	+268
Bud at The Glen	3	170	1	17	112	3	+326
Goodwrench 400	37	52	1	4	165	4	+213*
Goody's 500	3	170	1	Win	180	3	+203
Southern 500	2	175	1	7	151	2	+227
Miller 400	3	170	1	4	165	2	+232
SplitFire 500	2	175	1	Win	180	2	+227
Goody's 500	2	175	1	Win	185	2	+217
Tyson 400	7	151	1	4	160	2	+208
Mello Yello 500	3	170	1	37	57	2	+321
AC-Delco 500	Win	185	1	35	58	2	+448**
Slick 50 500	40	43	1	17	112	2	+379
Hooters 500	2	175	1	32	67	3	+487***

*Earnhardt led second-place Ernie Irvan by 27 points going into this race at Michigan, where Irvan was injured in practice, sidelined indefinitely, and Yates Racing withdrew.

**Earnhardt clinched his record-tying seventh championship in victory.

***This is Earnhardt's margin over Wallace. Mark Martin took second place in the final race and finished 444 points behind Earnhardt.

The Intimidator has his sights set on more championships, and passing King Petty. *Nigel Kinrade*

"Ernie was as consistent as Earnhardt until he was hurt," says Dan Rivard, first-year racing director of Ford Motor Company, "and I think he could have won. [He] certainly [would have] made the race closer. But far more important to us than a championship is that Ernie is alive and recovering miraculously. He has said his worst fear is being forgotten while he is sidelined. We have plans for him this year to make sure that doesn't happen."

"We don't feel badly about losing the driving title under the circumstances," says Morse. "Rusty made an outstanding contribution with the most wins and points toward the manufacturer's title his first year in Thunderbirds. We're proud of that and the effort of all the teams in Fords."

Earnhardt eventually won the title by a whopping 444 points over Mark Martin, who overtook Wallace by 43 points by virtue of his victory and Wallace's DNF in the season finale at Atlanta. Earnhardt had 20 top fives, 25 top 10s, two DNFs, and an average finish of 8.03.

Herb Fishel, director of the GM Motorsports Technology Group, is delighted with Earnhardt's and Chevrolet's season. "In a year when Fords clearly outnumbered Chevys on the race track, and the revised Thunderbird was thought to have a technical edge over the aging Lumina," says Fishel, "it's not a big surprise that Ford won the great majority of poles and races and the Manufacturer's Championship.

"But statistical performance characteristics are not always decisive in winning the major events, where superior efforts and extraordinary individual performances can overcome big odds.

"Sterling Marlin and the Morgan-McClure team proved that in the Daytona 500, and Jeff Gordon and the Hendrick Motorsports team in the Brickyard 400, just as they did in the Coca-Cola 600 with Ray Evernham's winning move of taking on two tires instead of four on the final pit stop.

"And Dale Earnhardt and the Richard Childress team proved all year long in winning their sixth championship together in the past nine years, and Dale's seventh, that people really make the difference in NASCAR Winston Cup racing."

Childress crew members, most of them together for the team's titles, say this championship is the sweetest because they knew how much it meant to their driver.

"Last year, our performance fell off at the end of the season, even though we were champions, because we were more conservative with setups than we needed to be," says Andy Petree, an undefeated champion in two years as crew chief. "This time we sat out and set up to win every race and we were consistently good.

"Dale was never better. He had to be aggressive to get to the front because we started 16th or lower 14 times, but he showcased his maturity by not increasing risks trying to stretch a second or third into a win. Our strength was consistency. That wins championships and Dale is the master." Earnhardt also won two poles, and the top winner in the Busch Clash shoots for his sixth victory in the dash for cash at Daytona on February 12, 1995.

"We were more intense than ever, especially in the second half of the season," says crewman David Smith, "and Dale took the intensity to a higher level because he knew he had perhaps his best opportunity to win [championship] number seven. Everything went in our favor."

"Each championship seems to get harder," says engine specialist Danny Lawrence. "But our goal is set on [championship] number eight."

Earnhardt commences on February 19, 1995, by trying for the 16th time to win the Daytona 500. That is another story indeed.

10

DAYTONA 500 SPEEDWEEKS
Earnhardt Denied Again

BY DON ALEXANDER
Circle Track June 1995

You're a crew chief for a top Winston Cup team. It's late in the Daytona 500. You're running in second. Your car is not handling well, and it looks very likely that your driver will fall back without a pit stop and a tire change. There's a yellow and no one else comes in, but you decide to make a stop. You go out in 14th place after the stop. Most crew chiefs would get walking papers for making this call, but you're Andy Petree and your driver is Dale Earnhardt.

It takes massive confidence in a driver to risk a stop, even during a caution, late in a race when no one else comes in. For Petree and Richard Childress, the decision was easy in this year's Daytona 500. Earnhardt is one of the few drivers who can get through traffic and actually have a shot at winning the race. And that is just what happened at the 500 in 1995. The pit stop, with 10 laps left and Earnhardt's drive through the field, made an excellent race a memorable event—one of the best in recent history. That Earnhardt fell short by one position is indicative of the outstanding performance of Sterling Marlin in the Morgan-McClure/Kodak Monte Carlo. The Tony Glover–led team fielded the car to beat, and no one else was up to the task. Had Earnhardt had half a dozen more laps, the outcome may have been different. But Dale didn't, and he fell short by one position in his 17th bid to win the most prestigious event on the Winston Cup calendar. It seems that Earnhardt is destined to be the bridesmaid at the Daytona 500.

But the points battle is a different story. Earnhardt is right where he wants to be, well ahead of his chief rival Rusty Wallace, who again started the season behind the eight ball after a crash left him with a 34th-place finish. Mark Martin looked strong, as did Dale Jarrett in the Robert Yates Ford, and Rick Hendrick's stable with Jeff Gordon, Terry Labonte, and Ken Schrader, who all have shots at the title.

But the two stories at Daytona were Earnhardt's move to the front after the pit stop and Marlin's dominance. How does Earnhardt slice through traffic so easily? Here are a couple of

Nigel Kinrade

possibilities. Earnhardt has a tremendous ability to judge speed and traction. He always knows exactly what his car can do, and can't do, at every instant. And he has the confidence to instantly execute what he has judged as the best move. Earnhardt has the ability to perceive what his car will do and what the other cars/drivers will do, and he uses that. The ability to perceive is nothing more than paying attention to what is going on around you. Earnhardt has taken that skill to a very high level. Everyone can improve their perception, but it takes considerable work to get to Earnhardt's level.

One of the tricks that Earnhardt uses to improve his perception is to expand his visual field. Broader and longer visual fields allow a driver to take in more data, to be more perceptive. It's a safe bet that Earnhardt looks farther down the track than anyone else, and takes in more information than anyone. Long visual fields allow a driver to improve perception and take a longer look into the

DAYTONA 500

Daytona International Speedway
February 19, 1995

1. Sterling Marlin, 200 laps
2. Dale Earnhardt, 200 laps
3. Mark Martin, 200 laps
4. Ted Musgrave, 200 laps
5. Dale Jarrett, 200 laps
6. Michael Waltip, 200 laps
7. Steve Grissom, 200 laps
8. Terry Labonte, 200 laps
9. Ken Schrader, 200 laps
10. Morgan Shepherd, 200 laps
11. Dick Trickle, 200 laps
12. Kyle Petty, 200 laps
13. Ricky Rudd, 200 laps
14. Lake Speed, 200 laps
15. Ward Burton, 200 laps
16. Ricky Craven, 200 laps
17. Loy Allen, 200 laps
18. Bobby Hamilton, 200 laps
19. Joe Ruttman, 200 laps
20. Geoff Bodine, 200 laps
21. Rick Mast, 200 laps
22. Jeff Gordon, 199 laps
23. Bill Elliott, 199 laps
24. Jeff Burton, 199 laps
25. Brett Bodine, 199 laps
26. Robert Pressley, 199 laps
27. John Andretti, 197 laps
28. Ben Hess, 196 laps
29. Randy LaJoie, 195 laps
30. Bobby Labonte, 185 laps
31. Derrike Cope, 184 laps
32. Darrell Waltrip, 180 laps
33. Davy Jones, 166 laps
34. Rusty Wallace, 158 laps
35. Jeremy Mayfield, 155 laps
36. Dave Marcis, 129 laps
37. Todd Bodine, 105 laps
38. Jeff Purvis, 57 laps
39. Mike Wallace, 57 laps
40. Steve Kinser, 27 laps
41. Phil Parsons, 27 laps
42. Joe Nemechek, 8 laps

Top 10 Winston Cup Points
February 19, 1995, Race #1

1. Sterling Marlin, 185 points
2. Dale Earnhardt, 175 points
3. Mark Martin, 170 points
4. Ted Musgrave, 160 points
5. Dale Jarrett, 155 points
6. Michael Waltrip, 155 points
7. Steve Grissom, 146 points
8. Terry Labonte, 142 points
9. Ken Schrader, 138 points
10. Morgan Shepherd, 134 points

MOVE OF THE RACE

Andy Petree's call to bring Dale Earnhardt in for a tire change during the last caution period, when no other leaders pitted.

future. This allows better, more accurate anticipation, and anticipation is a skill that Earnhardt uses with utmost accuracy.

The speed show by Marlin at the 500 was awesome. Some believe that more power was the key. Smokey Yunick has some ideas, and it certainly is possible that the Kodak engine had an advantage. The difference in noise was most likely an unusual header configuration. But the noise may have simply been a diversion from the real key to the No. 4 car's speed.

There are three ways to go faster on a race track: more power, more traction, and less drag. More power is the least likely reason for Marlin's success. More traction is also not highly likely. But less drag is very likely. It could be that the Morgan-McClure team had more success dialing-in the new Monte Carlo in race trim than the other teams. It is clear that the Monte Carlo is on par with or better than the T-Bird, and much better than the Lumina. So an aero advantage is a strong possibility. But tire drag is also a likely factor.

A tip-off that indicates that Marlin's car had less tire scrub was the fact that Marlin was running on the bottom of the track in the turns all race long. This requires a very good setup, perfect tire pressures, minimum toe, and all of the tires pointing straight ahead. Sticking in the low groove allowed the car to follow the shortest path around the track without losing speed. The low groove is always the fastest way on a banked track where the banking angle is consistent from top to bottom.

An additional factor is the driver. No matter how well the car is set up, the driver must minimize tire scrub by using the steering wheel as little as possible. The driver with the lowest-average steering angle per lap (or race) will scrub the tires the least and is in a strong position to win, just as Sterling Marlin was at the 1995 Daytona 500. Sterling drove a stellar race, and earned the victory along with Tony Glover and the entire Morgan-McClure team.

Earnhardt has the ability to perceive what his car will do and what the other cars/drivers will do, and he uses that. The ability to perceive is nothing more than paying attention to what is going on around you. Earnhardt has taken that skill to a very high level.

11

EARNHARDT ON PETTY & PETTY ON EARNHARDT

Stock Car Racing July 1995

Their names are linked together now, maybe forever, maybe for just this one magical stock car racing season. From the moment Dale Earnhardt clinched his seventh NASCAR Winston Cup Series championship last autumn, managing to not only touch Richard Petty's "untouchable" record but to wrap his hands around it, there has been no way to mention one without also mentioning the other.

The headline writers have had a field day, leaning on all of the appropriate and predictable clichés: A Pair of Sevens. Two of a Kind. And they committed what once would have been considered NASCAR blasphemy by elevating Earnhardt to Petty's level of royalty: A New King.

Petty and Earnhardt even got into the hype game themselves, in classic 1990s pro-sports fashion, by announcing their partnership in a souvenir line commemorating their shared record. The catch phrase? Seven and Seven.

In a career chock full of eye-popping wins and outrageous accomplishments, Dale Earnhardt had never managed to do anything quite this big. Matching Petty's mark—during an era even King Richard will acknowledge is vastly more competitive than his own heyday, the 1960s and 1970s—defined Earnhardt, once and for all, as the single most prolific Winston Cup racer of his generation. It was as if this one particular championship, number seven, turned him into a rock star overnight. He was everywhere you looked on television, both network and cable. He literally towered over New York City, appearing on the famed Sony Jumbotron screen in Times Square while in town for NASCAR's glitzy Waldorf-Astoria awards dinner. All winter long, he was the toast of motorsports America.

And yet in the midst of it all, Earnhardt—who is rarely loose with his compliments—said time and again, in so many words, "Aren't we forgetting somebody?"

"I may have won seven championships," Earnhardt declared to all who cornered him with a camera or a note pad, "but Richard Petty is still the King. He always will be."

Two veritable legends of racing. *Nigel Kinrade*

Earnhardt is still saying things like that, and his words warm the heart of his sport's monarch.

"I think he's been pretty sincere about it," Petty says, "and that makes me feel good. I don't think it's been, you know, just public relations."

Petty is sitting in his team's motor coach in a speedway infield, relaxing. He is wearing a smile, but it is not the famous mega watt grin the racing world saw in 200 victory lanes. This is just a small, honest smile, and Petty may not even know it is there, but it adds some weight to his words as he talks about how nice it has been to hear Dale Earnhardt, in his moment of highest glory, invoke the Petty name over and over and over again.

Despite his own successes, Richard Petty is not a showy man, not a braggart. He says, as if seven Winston Cup championships were no big deal, just one more statistic, "I look at it like, I've done my thing, and he's done his."

He pauses. "From Earnhardt's side of the deal, at least he's had someone to compare himself with. With the records I set, there was no comparison with anybody else. Anytime I broke records, most of the time the records were already mine. I just kept extending 'em.

"So it was hard for me back then to say, 'Wow, seven championships, that's great,' because I was only beating the guy who had won six, and that was me! It was the same way when I won the sixth one, and the fifth one. It was my own records I was breaking.

"Now, everybody's got something to compare Earnhardt's championships with."

But there is a difference, Dale Earnhardt says, in comparing records and comparing legacies. He is sprawled on a sofa in the lounge area of his team's transporter, speaking in the blunt tones he uses when he wants to make a point.

"Richard Petty is the King," Earnhardt says. "And I'm aware there's an opportunity out there for me to win eight championships, or more, because I still plan to race for several years. But if I do that, if I break our tie, I'll still feel like Richard is the King. He was the first guy to do this stuff, to set all these records. He set the pace, set the standard. He was a pioneer, a trendsetter.

"Richard did everything there was to do. He's in his rightful place when they call him the King, and I surely don't ever try to put myself above him."

I say to Earnhardt, "Don't you think it's sad that a lot of the youngest fans today, the TV fans, have really no idea of the impact that man had on this sport?"

He nods in agreement, and says sadly, "I don't think they do."

There is this long, thoughtful silence now from NASCAR's second seven-time champion. Then Dale Earnhardt says, "I don't say this to take away from anybody else, but if you had a book you could use to read back through all the years, all the drivers, Richard Petty's name would have to be right at the top."

They are products of different eras, as Petty points out: "I was successful in my time, Dale's been successful in his." Petty earned his titles in 1964, 1967, 1971, 1972, 1974, 1975, and 1979. In Richard's final championship season, Earnhardt was voted NASCAR's top rookie. One year later, in 1980, Dale won his first Winston Cup. It happened again in 1986, 1987, 1990, 1991, 1993, and 1994.

There were 30 years between Petty's first championship and Earnhardt's most recent, so they have essentially owned stock car racing for a generation. But there is, you must understand, a serious generation gap involved. Earnhardt's NASCAR is all grown up, with a 31-race schedule fueled by corporate dollars and television exposure. The sport has become respectable; you can see racing-related bumper stickers from midtown Manhattan to Manhattan, Kansas. Petty's NASCAR was more parochial; the big stock cars occasionally ventured out of the deep South, but they were never gone for long. Petty's NASCAR was busier, too. On the way to his first two championships, King Richard was as likely to be slamming through the ruts of the old South Carolina dirt tracks in Greenville and Columbia as he was to be gliding around Daytona or Atlanta. In 1964, young King Richard competed 61 times, winning nine events and beating, in order, Ned Jarrett, David Pearson, Billy Wade, and Jim Pardue in the point standings.

Petty raced in an era that rewarded brains. The Winston Cup cars of the day—then called Grand Nationals—were unrefined and prone to mechanical breakdowns. Having the fastest car in the field meant nothing if a guy wasn't smart enough to nurse it over the required distance. Petty Enterprises may have had the best equipment of its day, but it was also controlled by perhaps the most savvy driver in the business.

Conversely, Earnhardt's era—our era—rewards brawn. Next to Petty's old iron, today's NASCAR stockers are positively indestructible. Gone are the days of frequent blown engines and failed chassis parts. Almost everybody finishes. And because of the huge payrolls made possible by big sponsorships, there are more competitive teams today than at any previous time in the sport's history. It takes only logic to realize that if 15 or more fast cars are going to run on the lead lap until the finish, anybody who plans on winning will need to stand on the gas all afternoon long. And nobody stands on the gas like Earnhardt does.

Richard Petty has seen all of the changes. He says, "When I showed up, the equipment was not near as good as the equipment now. If Earnhardt had run the equipment we had the way he runs his stuff now, he would have fallen out of a lot of races."

"Would Dale have been a big winner in your day?" I ask.

"He would have been a big winner anytime," Petty says. "Oh, yeah."

"But wouldn't his style have hurt him a little bit?"

"It would've hurt his consistency, but it wouldn't have stopped him from winning races."

The King leans forward, as if for emphasis. He says, slowly, "If you took David Pearson, in his prime, and put him in one of these top cars today, David would win races. Cale Yarborough would win races. Bobby Allison would win races. And the people who are winning races today, like Earnhardt, would have won races back then, too.

"Winners are winners, and they would be winners in any era. The comparison I use is Johnny Weismuller, the guy who played Tarzan in the movies. He was a championship swimmer and won a bunch of gold medals. Now, if he swam a race in six minutes, today they swim it in four. But what you need to look at is, he was the best swimmer in his time, and if he came along right now, he'd still be one of the best, because he'd just train like they do today and take advantage of all the things the current swimmers use. He'd be just as good as the rest of 'em, because he had that determination, that spark, that a champion has to have.

Earnhardt: "I may have won seven championships, but Richard Petty is still king. He always will be." *Nigel Kinrade*

Earnhardt has been chasing Petty since 1979, when Petty won the championship and Earnhardt took home Rookie of the Year honors. *Staff photo*

"The good people, I believe, are just destined to be good. A champion will be a champion."

I say to Petty, "Do you ever wish you could have come along a little bit later than you did, so you could have run these bulletproof cars today?"

His answer catches me off guard.

"No," Richard Petty says. "My driving style was right for the time when I came along. I don't know that I'd fit in today as good as I did back then. Yeah, I feel like I could be competitive if I came along today. But, like I said, I think I came along at just the right time."

He looks out the window of his big luxury bus, toward the garage area. He knows Dale Earnhardt is out there, somewhere. For a moment, King Richard is silent. Finally, he says, "You know, destiny is a funny thing. There have always been certain people for certain times. Babe Ruth came along when baseball needed him, right after they had that Black Sox scandal. Things were looking pretty bad, and then here came Babe Ruth, hittin' 40 or 50 home runs a year, and people started really paying attention to baseball again. He was the right man to take that sport away from the scandal it had been through, and bring it into a different light.

"I feel like I had my time in stock car racing, and I helped expand it. Now it's Earnhardt's time to help expand it. Racing needs somebody like Earnhardt to carry the torch on to whoever the next cat is. Dale took it from where I carried it to, and he'll carry it a little bit further, and then somebody's gonna come along and take it even further, hopefully."

Maybe it is our good fortune that Petty and Earnhardt are from such different periods, such different disciplines. It allows each to look at the other from his own perspective: Petty, the master strategist, can see Earnhardt thinking. Earnhardt, the hard charger, has seen the gutsy, pure driver in Petty. In many ways, they know each other better than we ever will.

Listen to Earnhardt: "Richard could charge when he had to. He had several battles with David Pearson and Bobby Allison, and later on he had some hard races with Darrell Waltrip and the rest of us, too. I think Petty was just a great all-around driver. He knew how to make his equipment fast, and knew about his cars—that's what made him so good—but he was also a great driver."

Now, hear Petty: "The other day, somebody said, 'Man. Earnhardt's always putting out a hundred percent.' And I said, 'No, that's not right. Nobody ever puts out a hundred percent all the time.' Common sense will tell you that.

"See, a lot of the races I won, I'd just be out there running and running and running, and then when it came down to the end, somehow or other I'd figure out a way to win. That means I couldn't have been doing a hundred percent all day, right? And all these races you've seen Earnhardt winning right at the end, it's the same thing: if he was putting out a hundred percent all day, where'd the rest come from when he needed it?

"A race driver probably puts out 75, maybe 80 percent all day long. Maybe that exact percentage is wrong, I don't know; pick whatever number you want to. But, see, if everybody's putting out 75 percent, Earnhardt's 75 percent is better than anybody else's 75 percent. And when it's time to dig in, he's got plenty left.

"It's just a natural deal," Petty says. "You think you're doing everything you can, but you're not until you really need it. Earnhardt thinks, 'Man, I gotta get by this cat. There's only a few laps left.' And he draws out whatever he's got left, and he gets the job done."

Earnhardt wins, according to Petty, because he is a smart racer surrounded by a smart team.

"That part of racing hasn't changed," says the King. "The smart guys still wind up winning. There's a lot more good equipment today, but all that's done is put everybody else closer to the leaders. It used to be that we had four or five cars that were really good, and then a bunch of guys in the middle of the pack and a few in the back. Well, there's still the best four or five, but this other crowd has all caught up to 'em.

"When the year's over, there's still four or five teams that will win most of the races, and one of 'em will win the championship. That's the same as it's always been. The good equipment has tightened up the field, but it really hasn't made winners out of anybody. Maybe the guys in the back of the pack are on the same lap now, but they won't win races. It doesn't work that way."

Earnhardt, like Petty, believes that winning is a team thing.

"I'm a Richard Petty fan," Dale says, "but I'm also a big David Pearson fan, a big Cale Yarborough fan, a big Bobby Allison fan. I used to watch 'em all, and when I got started, I was able to race against 'em. Donnie Allison, Dave Marcis . . . there's so many guys who taught me things. And a lot of 'em, I think, were maybe as good as Richard Petty, but they just didn't have the right combination of teams and equipment that Richard had with Petty Enterprises. He was fortunate to be hooked up with the Chrysler Corporation in that era, because they were great back then."

So great that Petty and his family's team were annually the smart preseason pick to win the NASCAR championship. Just like a cat named Earnhardt and his car owner, Richard Childress, have been for, oh, the last 10 years.

Petty shrugs, says, "Right now it's Earnhardt, and then . . . everybody else. Earnhardt's team is a championship team, and there's no other championship team out there. There's winning teams out there, but there's not championship teams out there. I'm not saying that maybe one of those teams won't work into becoming a championship team, but right now, the best team is Earnhardt's.

"The biggest thing we have in common," Petty says, "is that Dale's had the Childress team, and I had Petty Enterprises, and those were the best teams for us. The driver was good—both me and Dale—but so was the team. Each one complemented the other. Neither one could have been as good without the other. I believe that.

"So the longevity we both had with our teams was smart. I never changed teams; basically, it was always me and Petty Enterprises. Well, once Earnhardt finally got settled with Childress and they started winning, they did the right thing by staying together. They'd have both been dumb, even if

Petty: "Put our old unpolished group on one side, and this new bunch on the other side, and Earnhardt would be right in the middle. He could fit in either way." *Harold Hinson*

they didn't like each other, to back away from a successful combination. That's just common sense.

"Look at Pearson," Petty says. "He'd win pretty good with somebody, but then he'd go and drive for somebody else, and it'd take a while before he got back to the top. If he had stayed in one place, he probably could have done as good as me, as far as records and things like that. [Pearson, a three-time NASCAR champion, had 105 wins, second only to Petty's record 200.]

"If Earnhardt had left Childress to drive somebody else's car, I think he might not have done as good as he has," Petty says. Then, laughing, the King adds, "And I know Richard Petty wouldn't have done near as good if he hadn't been driving for Petty Enterprises."

It is one of those questions we will never know the answer to, and yet it is the stuff of garage-area debate and race-shop speculation. How would Dale Earnhardt, the toughest guy on today's NASCAR block, have fared against Petty and his tough-guy rivals?

King Richard says, "Of all the people running today, Dale would have fit in with us better than any of 'em. He's the closest thing stock car racing has to a throwback. He's polished enough to run with this crowd today . . ."

. . . Petty stops, smiles . . .

" . . . but he's also unpolished enough that he would have fit right in with our crowd."

Now Richard Petty makes a sweeping arm gesture, meant to paint a broad stroke over the Winston Cup garage, and maybe the grandstands, too.

"This sport today," Petty says, "is a lot more polished than it was 20 or 25 years ago. Put our old unpolished group on one side, and this new bunch on the other side, and Earnhardt would be right in the middle. He could fit in either way.

"My son Kyle wouldn't have fit in with our crowd, I don't think. Rusty [Wallace] could maybe have come along with us a little bit. But Earnhardt, he would work both sides of that street."

Earnhardt agrees, which is not surprising. He is a man full of reverence for the old days, the old ways, and the age that produced his father, the late NASCAR sportsman champion Ralph Earnhardt.

"I think I'd have fit in," Dale says. "I'm pretty much from the old school. I've changed over time, and I think that's what's kept me competitive over the years: being able to change with the different procedures, the different rules, the different technologies of the cars and the tires, everything.

"But," he grins, "I do come from that old school."

The lessons he learned there, of course, got Earnhardt in plenty of hot water during his earliest days on the Winston Cup trail. His obvious aggression and his scrapes with other top drivers of the mid-1980s—most notably Darrell Waltrip, Bill Elliott, and Geoff Bodine—earned him NASCAR reprimands, as well as the nickname that has become his trademark: the Intimidator.

I ask Richard Petty, "The trouble Dale was getting into 10 years ago, and the way he drove . . . would he have gotten away with that stuff against you guys?"

The King gives me a shake of the royal head.

"No."

"Why not?"

"They'd have taken care of him," Petty says, leaving himself out of the equation. "What they would've done was, they'd have polished him up, one way or another. For Dale to survive, he'd have had to clean his act up a lot quicker.

"See, Pearson or Allison or Yarborough, they wouldn't have put up with his stuff. There was no way to intimidate those people. No way. He would have been wasting his time, OK? In fact, he would have been the one who got intimidated. People like them, they just wouldn't stand for that stuff."

Not that Richard Petty stood for much of it himself. Earnhardt recalls often being on the receiving end of some not-so-friendly advice from the kindly King.

"Oh, yeah," Earnhardt says, chuckling. "He's gotten out of his car and chewed my ass out a lot of times."

"Do you remember any incidents in particular?"

"All of 'em," Earnhardt replies. "There was Martinsville, when I ran over the top of him one time. There was Charlotte, probably in 1979, when I raced side by side with him and didn't realize Darrell Waltrip was pullin' away from us. Darrell won the race, Richard was second and I was third, and I was so excited about racin' with Richard Petty. Well, I got out of the car and I was in the locker room, changing clothes, and there was Petty. He said, 'Boy, if you'd have got in line with me, we'd have caught him, and me or you might have won this race instead of Waltrip.' I'll never forget that."

I say to Earnhardt, "When Richard Petty spoke to you in those days, was that like hearing the voice of God?"

Eyebrows raised, Earnhardt says, "Yeah. I mean, I was pretty intimidated by him. When he talked like that, if you were smart you listened and you learned. And I did."

Now they share records and headlines. And there are times, Dale Earnhardt admits, when the whole idea seems mind-boggling: his name and Richard Petty's name, together at the top of the NASCAR charts.

"It is pretty amazing," Earnhardt says. "I mean, I don't daydream about it; I don't sit and ponder on stuff like that. But, yeah, it's really something."

And what about that souvenir business? Who'd have ever imagined, back in 1979, that the wild young Winston Cup Rookie of the Year and the seven-time NASCAR champion would one day end up with their names and faces inked across the very same T-shirts and bumper stickers? Earnhardt says he's just grateful for the opportunity, however long it might last.

"The fans like something new," he shrugs, "and this souvenir thing is new, because of the seven championships. It's trendy. But I think after a while, it'll go away."

I tell him, "Aw, you're just hoping it'll go away because it'll be 'eight and seven' instead of 'seven and seven.' "

Earnhardt does not answer. He just laughs. To do anything more, to talk about the day when the championship record might be his and his alone, would be to tread on the King. And in Dale Earnhardt's world, you just don't do that.

12

INDEPENDENT MAKES GOOD

BY BOB MYERS
Circle Track December 1995

Is Earnhardt's Era Over?

Inside Talk—Richard Childress

Richard Childress ponders his future. Can he keep his team on top through the 1990s? *Nigel Kinrade*

Inside Racing: *What's the secret to your success, unrivaled in the past decade, as a Winston Cup car owner?*

Childress: People, good people who surround this organization, starting with Dale Earnhardt, and a list of others who are what I call "real racers" and have played major roles.

I think the difference between myself and some of the other car owners is I look at myself as a racer more than a businessman. There are not many real racers left. That is, people who want to win and do whatever it takes. I'd say only 30 percent of the drivers today are real racers. The rest are in it for the money—to make a good living and sit out there and run all day.

IR: *What's the key element in winning six championships in the past nine seasons?*

Childress: [There's not one element], but preparation, dedication, and determination are high on the list.

IR: *How do you stay on top?*

Childress: We have to stay up with the latest technology. We have one building devoted to engine research and development, and that's run by specialists. I bought two machines recently that cost a total of $800,000. It's expensive, but you can't win without it.

IR: *Are chances good this season for a seventh title for Richard Childress Racing and a record eighth for Dale Earnhardt?*

Childress: I think so. We had more wrecks in the first half than usual and they took their toll on points. But we're in good shape [in July].

IR: *Is the driver or the team more important in winning titles?*
Childress: There's a song about love and marriage that says you can't have one without the other. The driver is a member of the team and a team wins championships.

IR: *You were a driver. What separates Earnhardt from the rest?*
Childress: I think never wanting to lose at anything, regardless of what or where it is.

IR: *Your team has become the master of consistency. How do you achieve that?*
Childress: It's a plan or philosophy that goes back to when I was driving. I had to run for the championship every year, though I never won it, because I needed all the extra money I could earn to stay in business.

IR: *Earnhardt is still aggressive, but is he also a smarter driver than he was four years ago?*
Childress: Absolutely, and it shows.

IR: *Is Earnhardt the best driver now, and how would you rate him on the all-time list?*
Childress: I've been fortunate to work with him for 11 years, and some of the things he can do with a car still amaze me. I think he will go down in NASCAR history as one of the greatest.

IR: *You've also won the most races since, including your first title in 1986. Is your philosophy to go all-out to win races and let the points fall where they may, or to be more content with high finishes?*
Childress: Every race we've ever entered we thought we could win, no matter where we started or finished. We stick to the same place and plan. A lot of teams not running [not in contention] for the championship that go or blow late in the season make us look like we're not running as strong.

IR: *As a hands-on owner, how many days have you missed being at the race track with the team?*
Childress: Eight since 1973, even with broken feet and ribs when I was the driver.

IR: *Do you have a business relationship with Hendrick Motorsports, like a technical exchange?*
Childress: We work closely with Hendrick Motorsports on some R & D for the Chevrolet group, plus the Monte Carlo.

IR: *Do you think Earnhardt is bothered by the arrival of the new generation, led by Jeff Gordon and Bobby Labonte?*
Childress: I don't think he's bothered. He feels he has good years left and knows he has to race the young guys just like when *he* came into Winston Cup. If anything, they make him even more competitive.

IR: *How long is Earnhardt committed to you, and how long do you think he will drive?*
Childress: Through 1996, and we're negotiating now for 2000. He might drive longer than that.

IR: *How long is GM Goodwrench committed as primary sponsor, and what has having them meant to you?*

Childress: [GM Goodwrench is committed] through 1996 and negotiating. GM Goodwrench has helped make Dale Earnhardt a household name, our team a champion, and Winston Cup grow through national advertising and exposure.

The Childress team responds to adversity with a consistent tenacity few teams can sustain year after year. *Staff photo*

IR: *Who will be your driver when Earnhardt retires?*

Childress: No one will replace Dale Earnhardt as a driver. It's too early to think about who I'd get.

IR: *Who would you pick if he left now?*

Childress: That's a tough one. Suffice it to say there are several talented drivers, not all of them in Winston Cup, I'd look at.

IR: *What circumstances led you to quit driving and Earnhardt coming to you for 11 races in 1981?*

Childress: Financial and timing. I used to run in the top five, then fell to the top 10. Then I got pushed back to the top 15, and I wasn't content running there. I told a friend, Wes Beroth, that I was going to get out of my car in 1982 or 1983. Wes told Earnhardt, who was unhappy over Rod Osterlund's sudden sale of his championship team to Jim Stacy during the 1981 season. We got together. It was a matter of timing.

IR: *Were you planning to sell your equipment and disband your team before Earnhardt came along?*

Childress: No. I had planned to hire a driver. I was in the right place at the right time—and fortunate enough to get the reigning champion and Wrangler as a sponsor.

IR: *Why did Earnhardt leave at the end of the 1981 season?*

Childress: Simply put, he was too good a driver for my team at the time. One of four options he had was to drive for Bud Moore, and I recommended that he take it. By this time, we were in the same hunting club, and we stayed in close contact.

IR: *Didn't your team become well established in two seasons with Ricky Rudd as driver?*

Childress: Yes. It was a good combination for us and Ricky. We won our first Winston Cup race together [in 1983].

IR: *Why did Earnhardt return to your team in 1984?*

Childress: When he left we talked about getting back together. By 1984, I felt my team was a winner and good enough for him.

IR: *What's your relationship with Earnhardt?*

Childress: We have two: good friends—hunting and fishing buddies—and business. We try to keep them separate.

IR: *Would you have believed that your operation would gross more than $22 million in prize money?*

Childress: Never. But at the rate the sport's growing, hitting $50 million over the next several years wouldn't be a surprise.

IR: *What honor or award means most to you personally?*

Childress: The Bill France award of excellence I received in 1985. I've always wanted to put something back into the sport, not just take from it. The award, which came as a complete surprise, made me feel that I've done that.

IR: *Where were you born and where is your residence?*

Childress: I was born in Forsyth County, North Carolina, near Winston-Salem. I live now in Clemmons, North Carolina, about eight miles from the shops in the community of Welcome.

IR: *What did your father do?*

Childress: He repaired radios and made furniture polish, which he sold door-to-door.

IR: *Were you interested in cars at an early age?*

Childress: Yeah. Before I got my license, I'd drive cars and trucks around the farm.

IR: *What was your first street car?*

Childress: A '49 Oldsmobile.

IR: *How did you get interested in racing?*

Childress: My stepfather took me to a race at Bowman Gray Stadium, a quarter-mile paved track in Winston-Salem, to see guys like Curtis Turner, the Myers brothers, and Glen Wood. Then I went back and jumped the fence to get in and hang out.

IR: *What was your first association with racing?*

Childress: I started selling peanuts and souvenir programs at Bowman Gray Stadium when I was about 12.

IR: *When did you get your first race car, and where was your first race?*

Childress: A friend and I bought a '47 Plymouth taxi cab, and we flipped to see who would drive it. He drove it the first week and I [did] the second. I loved it so much [that] we bought a second car, a '54 Plymouth, and were a two-car team in three weeks.

IR: *When and how did you meet your wife, Judy, and when were you married?*

Childress: She was my high school sweetheart. We were married June 27, 1965.

IR: *Does she work in the organization?*

Childress: Yes, she handles the finances. She and our daughter, Tina, are key people in the company.

IR: *What do you do for recreation?*

Childress: Hunting is my getaway. I love it. I'm so busy that about the only way I can enjoy it now is to block out the time and go to Africa. A trip like that clears my mind better than anything I've tried.

WINSTON CUP ANALYSIS

By Robert Eckhardt

Earnhardt Starts Championship Drive

The Brickyard 400 marked the end of the second third of the Winston Cup season. On his arrival in the Hoosier state, Dale Earnhardt was third in the standings, trailing leader Jeff Gordon by 146 points. Earnhardt knew it was time "to go for it." The crafty veteran brought a brand-new car to Indianapolis. He started the race with only 40 laps on the car and proclaimed, "This is the best car we've had all season." If that is not enough to scare the competition, Earnhardt has two more new cars just like this winner back at the shop. Dale is now ready to battle the last third of the year for his eighth championship. With only 12 races remaining in the season, a number of teams, besides Dale's, are starting their drives for the championship.

Rusty Wallace had a spectacular run at Indy. He started in 24th place and finished in second. The Miller Genuine Draft driver had a real shot at beating Earnhardt until the last pit stop. Rusty had to almost stop while exiting the pits when two cars in front of him collided. That incident was enough to give Dale a good lead on the track. It was good to see the two black cars running first and second again and both going for the win. Points are important at this stage of the season, but the $566,000 first-place money was definitely a motivation for these two past champions.

Another great drive was put in by Dale Jarrett. He crossed the finish line in third place after starting 26th on the grid. Jarrett slowly moved up through the field while conserving his car. At the opening green flag his car was tight, and on each pit stop the crew loosened the car up until he had one of the fastest cars on the track. He needed 10 more laps to be a threat to the first two cars. In the last three races, the Havoline team has completely turned around a losing season. They have finished first, second, and now third, and they have moved from 16th to 12th in points. It looks as if all their bad luck is behind them, and they should be a force to contend with the rest of the way.

Earnhardt took home $565,000 and a big trophy for his win at the Brickyard 400 at Indianapolis in August. *Nigel Kinrade*

BRICKYARD 400

Indianapolis Motor Speedway
August 5, 1995

1. Dale Earnhardt, 160 laps
2. Rusty Wallace, 160 laps
3. Dale Jarrett, 160 laps
4. Bill Elliott, 160 laps
5. Mark Martin, 160 laps
6. Jeff Gordon, 160 laps
7. Sterling Marlin, 160 laps
8. Rick Mast, 160 laps
9. Bobby Labonte, 160 laps
10. Morgan Shepherd, 160 laps

The Intimidator inspects the result of some bump and grind racing. *Sam Sharpe*

IR: *You were in Africa when Earnhardt won his first race on a road course, this year at Sears Point. Will you, as he says, go back to Africa during the next Daytona 500?*

Childress: If I knew that's what it would take to win the Daytona 500, I would go to the North Pole. But I'm not. I want to be there.

IR: *Your team and Earnhardt win about everything at Daytona, so why not the 500?*

Childress: Circumstances. We've lost some ourselves and gotten outrun in others.

IR: *Why did you locate your business in rural Welcome, and can you give us some idea of how much the shops have been expanded?*

Childress: The land was available for expansion and the area is isolated from other racing teams. My guys don't jump from shop to shop. The first building in Welcome, built in 1986, is 20,000 square feet. Now we have seven buildings, including the museum, and more than 80,000 square feet. Considering our investment, we should win. There'd be something bad wrong if we didn't.

IR: *Looking at the sport, would you have predicted its growth?*

Childress: I always thought it would be big, but not as big as it is. I think it's a long way from reaching its peak.

IR: *What's the biggest growing pain?*

Childress: The bigger the sport gets, the more people are involved and the less personal it becomes. That's the one thing I don't like.

Nigel Kinrade

IR: *What's your view of the NASCAR Sears Craftsman SuperTruck Series?*

Childress: We were one of the first teams to get involved. I think the timing is good because of America's new love affair with trucks. There's a whole new fan-following out there. The series could help solve future Winston Cup expansion problems. It's too early to predict whether it will be able to stand alone.

IR: *What's it like to have Earnhardt as a competitor in the truck series?*

Childress: I don't ever focus on one team. His is just another vehicle to beat. We do kid each other occasionally.

IR: *Is there more cheating in Winston Cup now, or is NASCAR catching more violators?*
Childress: I think NASCAR is doing the best job in the history of the sport.

IR: *How is your car treated by NASCAR?*
Childress: It's always under a microscope from NASCAR competitors and fans.

IR: *Why do people cheat with so much at stake?*
Childress: Because of the intense pressure to win, which is harder than it has ever been.

IR: *Does cheating have a negative impact on the sport?*
Childress: I've never called it cheating. It's trying to get a competitive edge. It is very negative, especially for those who get caught.

IR: *What do you think of NASCAR?*
Childress: It's the greatest game in town. Being a family-owned business has made it a tremendous success. If not for NASCAR, I would probably be driving an 18-wheeler or selling peanuts at a local race track.

IR: *Where is Winston Cup headed?*
Childress: I can see it someday going into nearly all of the country's major markets and international.

IR: *You are 51. How many more years would you like to spend in racing?*
Childress: I'd like to be involved in racing as long as I'm breathing. By the time I'm 55, if I want to go to Montana for two or three days and show up on Sunday for the race, or spend time with my grandsons, I can do it. If I want to go to Africa, I don't want to feel guilty about missing a race like I did at Sears Point.

IR: *From what do you derive the most pleasure?*
Childress: In racing, walking into victory circle as the winner is hard to beat.

IR: *Anything you'd like to add?*
Childress: I'm a very lucky and blessed person. I've also worked hard all my life. The first money I earned was a nickel for chasing down a farmer's pig that had gotten loose when I was seven. I've been working ever since, and I think it has paid off.

13

DALE EARNHARDT

Stock Car Racing Spring 1996

Telling the Story the Way It Really Is

Harold Hinson

You have been everywhere, done everything, won everything. You are Dale Earnhardt, and you have been around the block. Which is why in the autumn of 1995, when all of stock car racing was hysterical over your NASCAR points battle with Jeff Gordon, you were the calmest guy in the garage area.

Your explanation is simple: "I've been down that road before."

Oh, yes. On seven occasions—1980, 1986, 1987, 1990, 1991, 1993, and 1994—you have been Winston Cup champion. Twice now, in 1989 and 1995, you have been the runner-up, and in 1988 you were third in the standings. Do the math: in the last 16 seasons, you have ranked third or better 10 times.

All of which made 1995 just another year at the office, in your case a hulking black Chevrolet owned by Richard Childress. Most drivers will never know how it feels to wake up on the last day of the season with a shot at the biggest prize in the sport; to you, the feeling has become almost routine.

You are Dale Earnhardt, and your accomplishments are so many, your trophy case so full, that you have grown immune to the hoopla. The only thing left that thrills you is the racing itself. Sure, the titles and the victories—68 by the end of 1995—are great, because they are a way to measure success. But racing isn't about glory: it is

"It's still exciting to get up on Sunday morning and get ready for a race. I guarantee you, there ain't another guy out there as excited about it as I am. There *can't* be. If there is, he's gonna have a heart attack." *Nigel Kinrade*

about sitting in the seat, sawing at the steering wheel. If racing is a drug, you are a classic addict: it somehow manages to both exhilarate and soothe you.

"It's still exciting to get up on Sunday morning and get ready for a race. I guarantee you, there ain't another guy out there as excited about it as I am. There *can't* be. If there is, he's gonna have a heart attack.

"And yet," you say, shrugging, "I'm calm about it, too. I'm excited, but I'm at peace with what I'm doing, and what's happening around me."

So exhilarating, and so soothing.

There is this physician who uses a little fingertip gadget to check the pulse rates of drivers prior to Winston Cup races. Invariably, yours is the lowest. You love that.

"Ain't nobody beat me yet," you say.

All those people who expected you to come unglued over Gordon and that title forget how you struggled through the sport's minor leagues on borrowed money and used parts. They don't know that the real pressure in racing is having just enough dough to pay the rent or the tire bill, but not both.

These days, you rest easy.

You say, "I know that on a given day, I could be the winner or I could be the loser. But I'm not gonna be panicking about it. Worrying just wears you out, gives you a few more gray hairs."

If anything turns you gray, it will be time, not trepidation. You will be 45 in April of 1996, and you have come a million miles from the dirt tracks where it all began for you. Who ever would have imagined that you, Dale Earnhardt, might one day be one of stock car racing's elder statesmen?

You would never describe yourself like that, of course, but in your own way you said as much recently when you allowed, "I reckon I'm comfortable with my place in racing."

A little fan appreciation before the 1995 Winston Select.
Sam Sharpe

And your place has a pretty good view. There are those who insist that among stock car racers, you are alone at the top, in a class of your own. It is high praise, and you do your best to duck it: "I don't believe that. I think there's more than me."

But you will agree that you have never driven better than you are driving right now. The aggression that was your signature in the 1980s is still there, you say, but it has been tempered by wisdom.

"The talent is the same," you say. "I think I've matured. I'm more calculating, probably, than I was then. But as far as being able to finesse a car, that's the same."

You frown, a bad memory surfacing. "Darrell Waltrip said something back in, like, 1986, 1987," you recall. "He said, 'Earnhardt, he ain't got no finesse.' But that's how I've won a lot of my races, by being able to finesse that race car."

Well, give Waltrip the benefit of the doubt; the two of you simply define finesse differently. Waltrip runs a smooth race, staying well within the boundaries of control until late in the going. You flirt with those boundaries all day long, drive on the ragged edge, and rise to meet any challenge that anyone—*anyone!*—dares offer.

From the beginning, you say, "There was a fire inside me. If you'd go *this* deep, I'd go *that* deep. Whatever it took to beat you, that's what I was gonna do."

It worked. You have beaten all the best drivers, whoever they happened to be at the time. There were, you recall, "Cale Yarborough, Waltrip, the Allisons, Petty, Pearson; then it was the Rusty Wallaces and the Bill Elliotts and the Terry Labontes; now, here comes the new guys, Bobby Labonte and Jeffrey Gordon and the Burtons."

In 1977, your sponsors weren't much, just a local speed shop and an engine builder. Now, corporate giants fall over each other to be involved with you. *Courtesy Lowes Motorspeedway*

You sized up everything you had to face along the way—new drivers, new tracks, new rules—and you overcame it. You say you wouldn't be a seven-time champion "if I had stayed set in my ways." You adapted, and you became better, and the gap between you and your competition, a gap spoken of freely in the garage area even if you won't acknowledge it, just continued to grow.

Jeff Gordon describes you as "probably the best driver I'll ever race against." Mark Martin has called you "the best race car driver in the business." Rusty Wallace, your good friend and fierce rival, labels you "the very best driver in the country," and says you have "more natural talent and more guts than anybody."

Wallace does not stop there. Your inventory of tricks, he says, is "incredible." He is absolutely right. You could market a highlight film of your most memorable moves, the wild passes and the spectacular saves, the jukes, the swerves, and the slides. You do things that other drivers cannot do, or will not do, and you have more last-lap secrets than anybody in the game. "There's things that happen through your career that you lose by," you say, "and you know not to lose by 'em again."

Maybe the difference between you and everybody else is this: even stronger than your desire to win is your desire not to lose. "It's not in the plan to go out there and lose," you say. "You go out to win."

You will concede that "you can't win all the races. You can't win all the championships." But you are Dale Earnhardt, so you can win more than mere mortals can.

A while back somebody cut through all the bull and just asked you the question straight out: "What makes you so good?"

You responded, "I'm dedicated to what I do. I'm not here to be flashy. I'm here to win races. I've worked hard to be a good racer. Where that takes you, you just don't know. Hopefully, it takes you in the right direction, and with me it has."

Yeah, you might say that. You are NASCAR's Michael Jordan, the sport's biggest star. Like you, Richard Petty won seven Winston Cup championships, but he had the ill luck to hit his peak before racing approached the near-mainstream status it enjoys today. Therefore, Petty never got the chance to be the one-man merchandising bonanza you have become in the 1990s.

He's your car owner now, and together you have won six Winston Cup championships. But in this photo you are on the inside in an unsponsored Winston Cup car in 1979, the year you won Rookie of the Year. In 1980, you'd win the Winston Cup for car owner Rod Osterlund. All the rest of the championships have been with Childress Racing. *Staff photo*

Among real race fans, an Earnhardt T-shirt means as much as those replica Jordan jerseys mean to playground kids. You are Dale Earnhardt, who came out of the North Carolina mill town of Kannapolis to become an icon. Your name, your car number, and assorted other logos are trademarked in so many places worldwide that you've actually lost count. "What is it," you ask your business manager, Don Hawk, "12 countries now?"

You travel in your own Learjet, relax occasionally on your own yacht—named *Sunday Money*, because that's what paid for it—and earn millions of dollars annually from souvenir sales, from your chicken and cattle farms, and from your Chevrolet dealership. Yet those who know you swear you are telling the truth when you say, "The racing comes first."

Money was never a goal for you, just a reward. You would race as fiercely for $5.65 as for the $565,000 you earned in winning the Brickyard 400 at Indianapolis last August.

You wave off your status by saying, "I'm still a regular guy. I love to wear jeans, ride in my pick up truck, have a beer once in a while, and go about my daily chores." But fame has rearranged your life in so many ways. For one, your words now carry immense weight. Funny, isn't it? Not long ago, you were "in awe of just being in

Along the way there have been some wild rides. This one at Pocono in Bud Moore's car was one of the worst. *Staff photo*

"My career is just getting started." *Sam Sharpe*

Your best move—marrying Tommy Houston's pretty niece. *Sam Sharpe*

Winston Cup racing," and today the press asks your opinion on every move NASCAR makes. Funny, too, how easily your answers come; once you worried only about the next race, and today you look way down the road, wondering where your sport will end up.

For instance, you say, "It's inevitable, I think, that at the rate we're growing, we're going to die out at some places." You mean the old half-mile tracks, which cannot draw the crowds, and thus cannot pay the purses, that the bigger superspeedways can. You foresee a schedule involving fewer short-track events.

"I'm not particularly for that," you say, because you love the short tracks. "But it's inevitable that it's gonna happen."

You see NASCAR embracing several new tracks popping up from coast to coast. "We need to be in [Las] Vegas. We need to be in Texas. We need to be in Miami. We need to be in California. Our sponsors need it, and that will drive us to go there."

Your opinions on the business side of the sport are noteworthy, but it is more fun to hear you just talk racing. Take your view of last season's Chevrolet vs. Ford rules controversy, and NASCAR's efforts to help the Thunderbirds keep up with the Monte Carlos. The cars, you said, are just one reason some teams win, and not the only reason. When NASCAR impounded your

Chevy and a top Ford for wind-tunnel tests after the final race, you could not let it pass without comment.

"They should have left the cars at the track, run them again the next day . . ."

You paused for dramatic effect, and then delivered the zinger: "And just swapped drivers."

You are Dale Earnhardt, and you are a proud man. There were times in 1995—the Year of Jeff Gordon—when the press treated you as little more than a supporting actor, and that didn't sit well with you. When the media's interest returned after your late-season charge, you made your feelings clear. "At one time this year, you guys weren't even worried about me," you told a group of writers. "Why were y'all not worried about me? It's like I was written off."

You cited the second Brickyard 400 as an example. "Media-wise, the focus wasn't on anybody but Gordon. Fairly so, because he was the winner of the first one, and he had fast time [for the 1995 race]. So it was Gordon's week. But they forgot about the race."

Which, of course, you won.

Even now, too many damn writers forget that you had a great year, winning five races. All they want to ask you about, it seems, is Gordon and his championship. They want you to compare him to a young Dale Earnhardt, which is crazy. Jeff Gordon, all of 24 now, has had a career so meteoric that he truly deserves the "Wonder Boy" nickname you hung on him. Your early days, on the other hand, were a monumental struggle: your father, 1956 NASCAR National Sportsman champion Ralph Earnhardt, was still showing you the ropes of the sport when he died in 1973. You were 22 years old, a high school dropout, married for the second time, and broke.

Gordon climbed a ladder. You shinnied up a greased pole.

"When I was 24," you recall, "I was starving to death."

You were 27 by the time you got a full-time Winston Cup ride, 28 when you were named Rookie of the Year in 1979, and 29 when you clinched your first title the following year. The sudden switch from obscurity to celebrity overwhelmed you.

"I had great friends like [journalist and publicist] Joe Whitlock to help me through all of it," you say. "But still, it was a big change."

Gordon, you believe, is "taking it all in" better than you did. "He's been schooled pretty good. He's been under the media blitz, and under the sponsor blitz, from the start."

Now he has been under the Earnhardt blitz, too. You knew he hated all that Wonder Boy stuff, which is exactly why you kept it up, and you couldn't resist announcing that you were the "first man to win the Brickyard 400." But in your unguarded moments, you appear to genuinely like the kid. You remind people that Gordon's title was no fluke. "He raced, and he won it. Nobody *gave* it to him."

Especially you. Down the stretch, you threw everything but the steering wheel at Gordon and his team. "We just came up short," you say. But you only missed by 35 points, after having been behind by more than 300 with six races to go. You proved, again, that even when you are down, you are never out. You admit to starting the season finale at Atlanta with one thought in mind: "Let's *bury* these guys." It was a message to everyone who had written you off.

You grin. "It ain't often you have a race car like that. I kicked some ass, and I had a big time doing it."

Showing up Gordon was never on your mind at Atlanta, you claim, but you do relish this: "I stole a little of his thunder that day."

That is a proud man talking.

You were no less proud when you hobbled to the ambulance after crashing at Michigan in June. It was the hardest hit you had taken in years, but you weren't about to show us anything more than a

"It's not in the plan to go out there and lose. You go out to win." *Harold Hinson*

limp. Not until the season ended did you let on, "I played hurt after Michigan. Had a broke back. Cracked a vertebra, crushed a disk. I can still move like this"—you flex your back, wince involuntarily—"and it hurts. For a long time, I couldn't drive good."

Now you stiffen, wary of sounding like a crybaby, which to you would be the worst of crimes. "That's not an excuse," you say pointedly.

You are asked what seems like a common-sense question: "Why didn't more people know about your back during the year?"

"Because I didn't tell anybody," you roar. "I didn't say anything. I didn't go to the doctor for four weeks."

Too tough. Too proud.

Pride is a constant theme with you. Your driving vision is eagle-sharp, but just last year, you realized that your eyes needed a little bit of help for up-close work. You bought a set of reading glasses, the same kind so many folks your age have to wear. But would you dare take them out of your pocket at the race track?

"Noooo."

You are your father's son. You believe that if you inherited anything from Ralph Earnhardt, it is "his determination, his savvy to take something that was lesser than the other guy had, and win with it."

Actually, Ralph passed down more than that. You described him as "an independent cuss," and in your case the apple did not fall far from the tree. In an age in which so many big-league drivers are PR-conscious good guys cut from the same bland cloth, you stand out. You don't always say the right thing. You don't smile when you feel like scowling. You are Dale Earnhardt, and you live by your own code.

Like the beer ads say, you are your own dog. Always have been. NASCAR spent much of the 1980s trying to smooth your rough edges, slapping your wrist every time you wrinkled a fender. All of the sanctioning body's tough talk went in one of your ears and out the other. Eventually you smoothed out the rough edges yourself, but the point was not lost: you weren't about to let anyone tell you how to drive a race car.

Driving styles, you have always maintained, are a personal thing. Because of your success, the media always wants to dissect your style, and you hate that. "People say, 'How do you do this, how do you do that?' I don't analyze it."

A reporter once asked you to discuss how you developed in your first few years as a Winston Cup racer. Your answer was pure Earnhardt: "I don't know; I drove my ass off."

Sometimes, the press has a hard time with independent cusses. Sometimes, independent cusses have a hard time with the press. You have always been described, accurately or not, as a tough interview.

Asked to describe your relationship with the media, you say, "I try to do the best I can with 'em."

"OK," the questioner says, "but are you happy with the relationship?"

"If I was sad with it," you reply, "what good would it do me?"

You are, essentially, a shy person forced by your job to live a public life. What's amazing is how you have managed to pull it off on your own terms. A long time ago, you drew a line in the sand, with your private life on one side. Nobody, the press included, can cross that line without being invited. Even then, you declare, "I'm gonna say what I want to say, and keep what I need to keep inside."

Good for you, Dale Earnhardt.

When you do allow the outside world to peek into your life, another Dale Earnhardt emerges. Here you are thumbing through a copy of a racing magazine, having promised to show somebody a photo of what you call "the best move I ever made." You find the right page, smile, and hold it up for display. It is a photograph of the former Teresa Houston, to whom you have been wed since 1982. It is no accident, you say, that when she showed up, your life and your luck improved.

You describe yourself today as "happy. *Very* happy."

Lord knows, it is a happiness you have earned. You were married twice before your 25th birthday, back in the lean times, and both unions were short-lived. Those marriages produced three children, and some leftover sadness shows on your face when you say, "I wasn't a good dad when they were young."

Those kids turned out fine, but you make it sound as if this happened in spite of you, rather than because of you. Still, you are now close to your sons Kerry and Dale Jr. (26 and 21, respectively) and your daughter Kelley (23), and they have taken up the Earnhardt family business: today, with your help and your blessing, all three are racing on Carolina short tracks.

Your youngest child, six-year-old Taylor, is the apple of your eye. There are times when she and Teresa meet you in Victory Lane, and your face lights up, and you don't look anything like

You recently signed a new long-term contract with car owner Richard Childress. You'll be winning races together into the next century. *Harold Hinson*

You have beaten all the best drivers, whoever they happened to be at the time. This time, it's Cale Yarborough feeling your front bumper. *Staff photo*

After being named NASCAR's Winston Cup Rookie of the Year for 1979, you started the 1980 season off right by winning the Busch Clash and its $50,000 first place money. *Staff photo*

the tough guy you are supposed to be.

You are Dale Earnhardt, and after too many miles of bumpy roads and too many years spinning your wheels, you finally know what it is to be content.

When Wrangler sponsored you, their advertising campaign called you "One Tough Customer," which indeed you were. Pairing the jeans with the sport's toughest driver was brilliant sports marketing. *Staff photo*

But it would be a mistake, you insist, to confuse contentment with complacency. People hear you talk about a peace in your life, and about being comfortable with your place in the sport, and they wonder if this is the language of a superstar edging toward the door. And you say, in so many words, "Not a chance."

"My career is just getting stared," you proclaim.

You are smiling, but you aren't kidding.

"I'm just getting good at this. I'm really starting to get a feel for it. I think I can do this better."

Honest?

"Honest," you say. "I'm learning how to do this. People talk about, 'Well, he's 44, and he's looking to slow down . . .' "

You frown. Then you mutter a couple of very bad words.

You have a plan for 1996, a mission. It is the mission of a man who wants to ensure that it will be years before anyone writes him off again.

"I am going to win races," you say, "and I am going to win that championship. You can count on that."

Now you narrow your eyes, and you say again, quietly, "You can count on that."

And who would bet against you? You are, after all, Dale Earnhardt.

14

BY RAY COOPER
Circle Track February 1997

INSIDE TALK—
Dale Earnhardt: All Business

Nigel Kinrade

Dale Earnhardt's accident at Talladega Superspeedway on July 28, 1996, in the DieHard 500 put the seven-time NASCAR Winston Cup champion in an unfamiliar position for the remainder of the 1996 season. He entered Talladega just 12 points behind points leader Terry Labonte, but on lap 117 of the rain-shortened 129-lap event, Earnhardt's Chevy was hit from behind when Ernie Irvan bounced into Sterling Marlin.

When the smoke finally settled, Earnhardt had sustained serious injuries for the first time since 1982. A fractured sternum and collarbone forced him to get out of the car during the first caution the next week in the Brickyard 400 at Indianapolis Motor Speedway and the rest of the season was an uphill battle for the driver of the Richard Childress No. 3 GM Goodwrench Service Chevrolet.

Earnhardt won two of the first four races in 1996 and was locked in a tight four-way battle with defending series champion Jeff Gordon, Labonte, and Dale Jarrett prior to the accident. After pulling one of the biggest upsets of the season by winning the pole and finishing sixth at Watkins Glen just two weeks after his accident at Talladega, it appeared Earnhardt might overcome his misfortune. David Green, standing by as relief driver for Earnhardt, never got in the car at Watkins Glen. Earnhardt admitted he had a car capable of winning the race and said he gave

Earnhardt says racing gave him everything good in his life. He shares his love of racing with his daughter, Taylor, by bringing her to most of the races. *Nigel Kinrade*

out toward the end, but no one questioned his determination after the gutsy performance.

No one has ever questioned the Intimidator's grit, but the rest of the season didn't go as planned, and he failed to win his record eighth championship. Earnhardt will turn 45 years old on April 29, 1997. The Kannapolis, North Carolina, native says he isn't ready for the rocking chair and plans to keep on racing as long as he feels competitive. When he isn't racing, Earnhardt stays busy working on his farm, helping his three children with their racing careers, attending to business matters with Dale Earnhardt, Inc., and planning for the future when he finally crawls out of his stock car for good.

What Earnhardt discovered shortly after the accident at Talladega was that climbing out of the car won't be easy.

Racing Warrior

Circle Track: *How tough was it for you to have a relief driver at the Brickyard 400?*

Earnhardt: There's only one thing that's been harder for me, and that was going to Neil Bonnett's house to see Susan, David, and Kristen after he was killed in that crash. That's completely different from me not getting to race. That's a very different feeling, a different hurt. That was hurting for someone you loved.

CT: *You got pretty emotional after getting out of the car.*

Earnhardt: I think that made people see I'm human, that I know how it feels to hurt. I don't mean from the injury, I mean from the heart. I'm not embarrassed by what happened. It's the way I felt. You didn't see one-tenth of how I was really feeling inside and just how close I came to bawling like a baby.

Earnhardt knows his racecars, and gets involved when he chooses to. *Staff photo*

CT: *Did the accident make you think about what racing really meant to you?*

Earnhardt: I told [wife] Teresa that everything I have has been given me through the grace of The Lord, but that I also knew that I had it because of racing. That includes her because I met her through racing, at a race track. If I wasn't racing, I wouldn't have met her. Everything I own and have the good fortune to have has come through racing, and I'm thankful for that. That's how good racing has been to me, so that's why it's hard for me to get out of the car, to give it up. Racing is an extension of me.

CT: *You were leading the race at Talladega when the accident happened. What do you remember about it?*

Earnhardt: When the car turned abruptly sideways I knew I was going to hit the wall. When it hit the wall is when I broke my sternum. When the car got on the side and got up in the air a little bit it was spinning around. I saw a flash and another car hit me at the same time. It was probably Derrike [Cope]. There was a big crash and the car went airborne again. That's when I broke my collarbone and bruised my pelvis. That's when the No. 33 car hit me. The car sat back down on the ground on its wheels, spinning around. The No. 29 car went by to my left and then a red car hit the front end. I assumed at that point, from the replay I couldn't tell much, it was either Schrader or the No. 94. Come to find out, it was Schrader. He said, "I seen you and I aimed for you."

The car stopped and the smoke was running out from under the dash, because the wires were burning because the dash was knocked down in the car so far from the top collision. I switched the battery switch off and I started trying to unbuckle my helmet. I knew my collarbone was hurt. About the same time the safety crews and Steve Peterson from NASCAR and

Earnhardt enjoys the outdoors and family life, but says he's "pretty relaxed around a racetrack." *Nigel Kinrade*

Buster Alton got to me. I told 'em not to cut the top off because I thought I could get out. They worked with me and pulled me out of the car. I wanted to lay down. I didn't want to stand up, but I couldn't because it hurt too bad to lay down. My chest hurt too bad so I said, "Just walk me to the ambulance." That's why I was walking instead of laying on the stretcher. I didn't want to walk.

CT: *How do you feel now?*

Earnhardt: I'm still sore every now and then but I feel good in the car. I don't want to hit too hard. I don't want to hit at all, as a matter of fact.

The Outdoors

CT: *Racing is your life, but away from the track, you love the outdoor life. Why?*

Earnhardt: I grew up around my uncles and my dad and we were always going bird hunting or rabbit hunting or squirrel hunting or fishing. I think that was the thing to do on Saturdays when people grew up back in those days. Instead of going to golf or play tennis, they were off rabbit hunting or fishing. It's not like that anymore. I wish it was. The wilderness has become towns. You've got to either have your own place to hunt or really plan a hunting trip these days.

CT: *You love your farm, too. What's it like?*

Earnhardt: We have cattle and Perdue chicken houses. We produce eggs for Perdue farms. We have some wildlife on there, but we mainly just enjoy the property. I'm lucky to get one day a week on the farm. I get up early every morning whether I'm on the farm or not. When it's daylight, I wake up. On the roadside property we have the Winston Cup shop going up. The Busch and truck shop is there and also Dale Earnhardt, Inc., business offices are there.

CT: *Do you have a favorite animal?*

Earnhardt: I like my horses. We've got about 14 horses, and Teresa and [daughter] Taylor and I ride 'em. I really enjoy that.

CT: *Do you operate tractors or heavy equipment?*

Earnhardt: Yep. I have my own bulldozer and my own motor grader, my own backhoe and tractors, and mini bulldozers. We have hay-binding equipment. I love making hay and I love to bulldoze things. I don't push trees down unnecessarily. I think it's wrong to cut down trees unless there's a problem. We had pine beetles in a lot of pine trees on our farm and we had to do a lot of cutting there.

CT: *What do you do to relax?*

Earnhardt: I relax just doing nothing. I'm pretty relaxed around a race track. I enjoy the racing, but when the racing is over I enjoy being home, too. We've got a lot of things going on back home, and I enjoy being back there.

CT: *Do you prefer the simple life?*

Earnhardt: Basically, yes. I enjoy going to the islands and the Bahamas and stuff. I like going to the simple islands. I don't like going where there's a lot of people. I like going to an island that has a small marina on it where you can get away and there's nobody around. I just sort of like the simple things.

CT: *Could you live in the house where you were raised and be content?*

Earnhardt: Sure.

CT: *Do you have a tree stand on the farm where you sometimes go to meditate?*

Earnhardt: I meditate all the time. I meditate when I'm on the airplane, when I'm on the ground, wherever.

CT: *How big is the stocked lake on your farm and do you enjoy fishing in it?*

Earnhardt: It's a 10-acre lake, and there's plenty of fish in it. Large-mouth bass and bream and catfish. I eat some of the catfish, but I put most of 'em back. I need to keep more of them because if it gets overpopulated, the fish won't grow.

CT: *Where are your favorite places to hunt?*

Earnhardt: I enjoy New Mexico. I go out there elk hunting and go up on the mountains. That's a great place to go and a good place to get away. It's a long ride in by horse getting to camp and a long ride to where we hunt. You're sort of one-on-one with nature there. I have quite a few hunting trophies.

CT: *How do you feel about anti-hunting groups?*

Earnhardt: Everybody's got their own rights to their opinion. If you do anything extreme, I think it's wrong. I've killed three deer this year, and I've killed one big deer. I pass up more animals

According to Dale, his wife, Teresa, is the "backbone" of his business. She announced that Modified driver Steve Park would drive her Busch Chevrolets in '97, replacing Jeff Green. Could the Earnhardts be grooming Park to pilot their Winston Cup ride in '98 or 1999? *Nigel Kinrade*

than I used to. I look at the quality of the animal now rather than if it's just a deer. Also, I've hunted a lot of areas where the farmers are getting eat up by the deer. They're overpopulated and cars are hitting them. They're a nuisance. I think if you'll ask North Carolina biologists or most biologists, their problem with the deer population is there's too many. It shows if you don't control the numbers, you overpopulate and the deer will become smaller and die of disease. If they die of disease, it creates more problems.

Family Business

CT: *As busy as you are, how do you maintain control of your businesses?*

Earnhardt: You hire good people. It takes a lot to manage my time, more so than it does my company. Teresa helps me manage my time. Dale Earnhardt, Inc., takes up most of my time outside of racing. Sports Image and the dealership takes up a lot of time. I try to go to the dealership but it's hard.

CT: *How would you respond to a report that you make five times as much money off the track as on?*

Earnhardt: No comment.

CT: *How much is your wife, Teresa, involved in the business?*

Earnhardt: Very much. She's really the backbone of the business. When it comes to the major decisions, her input is very valuable because of her education and common sense.

CT: *How did you and Teresa meet?*

Earnhardt: Her father introduced me to her at the race track. It was at Hickory, North Carolina, and she was 16. It was a long, long time before we started dating. We got married in 1982, and we met in 1974.

CT: *What do you and Teresa and Taylor enjoy doing together?*

Earnhardt: Business takes too much time away from us, but we take Taylor with us most of the time. Some tracks don't have real good facilities and we don't take her; she stays with her grandparents or she'll stay with my sister. She can go from one extreme to another. She can be a prim and proper little lady or she can be quiet and well mannered or she can be a regular little girl, too. She's a race fan, but she's a having-a-good-time fan. She enjoys the outdoors and it's hard to keep her inside. She likes to go outside like her daddy.

CT: *How do you feel about your other three children racing?*

Earnhardt: Dale Jr. is making some progress. Kerry and Kelley have had some disappointing runs

Steve Park meets the press to answer questions about his ride with Earnhardt Inc. *Harold Hinson*

this year. It's there for them if they want to do it. They've got to start showing some progress or it's going to be hard to keep doing it. They've all got their education and got jobs and they're doing well. If they want to race, they need to focus on it a little more and make it positive.

CT. *Why did you quit school at age 16?*

Earnhardt: Dumb. I just didn't see the opportunity. Why does anybody do anything when they're 16? It was really a big misfortune that I didn't finish. It was a big mistake in my life. You can give advice about staying in school, but all you can do is give advice. Your dad gave advice and my dad gave advice. We didn't always listen to it.

CT: *Did your father inspire you to be a race car driver?*

Earnhardt: All I ever thought about was racing. That's all I grew up knowing and that's all I wanted to do. My dad saw me race and he knew I was going to race.

CT: *What was it like growing up in the Earnhardt household?*

Earnhardt: I thought we were rich. We never wanted for anything and there was always something to do. I really missed it if we didn't get to go to the races. My mother still lives in Kannapolis, but she doesn't go to the races much anymore.

The Intimidator's Interior

CT: *Do you ever sit back and admire the empire you've created?*

Earnhardt: No. I walk in and admire Childress' shop and the cars and look around at the things that are going on for me. It's still hard to believe the things that are going on for the driver. I'm proud of what Teresa and I have in life.

CT: *Can you be yourself at the race track?*

Earnhardt: Yeah, if people will leave me alone, but that's the nature of the beast. It's just part of it. You have to take it and go.

CT: *Do you feel you have to live up to an image?*

Earnhardt: No, I just try to be myself and let things take care of themselves.

CT: *Do most people know the real Dale Earnhardt?*

Earnhardt: No, but the majority does. It's not a worry of mine if they do or don't. I just try to be myself and go about my business.

CT: *Are you shy to a certain extent?*

Earnhardt: Not as much as I used to be.

CT: *Would you rather be on your farm than in the public spotlight?*

Earnhardt: Hopefully, I'll own my own team some day and be like Richard Childress, but I'm going to race as long as I'm competitive and feel good. I'm excited about the '97 Busch team. I think Steve Park is going to be a real good addition to our team. He's at a point right now where Gordon was when he got into Busch Grand National. It's taking a bit of a chance on the experience side of it, but still, I think the drive I see in him will take up the slack and make up the difference. That's how Gordon was born. He got into Busch and went on to Cup.

Hopefully, I can take him where he needs to go with Dale Earnhardt, Inc. I've always got Childress to fall back on, too. If things progress like I think and like I hope, I can see him in Winston Cup in '99 or earlier. I like him because he's a talent, and I like the way he drives. He's smooth, laid-back with what he wants to do. He talks about his dad and it sounds like when me and my dad raced. I grew up around my dad and there's a lot of parallels there. He doesn't have any hobbies. He doesn't hunt and fish. That race car is what's important to him, and that's great. That's the way it's always been for me.

15

LARGER THAN LIFE

Stock Car Racing March 1997

By "Giving 'Em What They Come to See," Dale Earnhardt Has Become the Biggest Name in Stock Car Racing

Ralph Dale Earnhardt, who as a kid had little patience for teachers and less for the stuff they taught, cut class for good long before the North Carolina school system had a chance to get any John Steinbeck into him. And you can be assured that this unfamiliarity went both ways: the novelist died in 1968, long before Earnhardt had amounted to anything. The record shows that neither man suffered for this mutual ignorance. Earnhardt, who wanted nothing more than to be a stock car racer, is a seven-time NASCAR winner.

This is the kind of thing that Earnhardt himself would never suggest, because it risks introspection and an analysis of his driving style, a process he loathes. So his praises are best sung by others, and in their efforts to describe the man he has become, on and off the track, they end up quoting everyone from Rusty Wallace to, well, John Steinbeck.

Surely it all must get a little highbrow at times for Earnhardt, who has never pretended to be anything more than a Carolina boy who worked as hard as he drove and became a star as a result. Forget Steinbeck; he is more comfortable with the literary works of his friends Kix Brooks and Ronnie Dunn, the balladeers who form the country duo Brooks & Dunn. But even they cannot resist the challenge of trying to sum up the Earnhardt phenomenon: in the acknowledgments on the jacket of a recent album, the boys tipped their hats to "Dale Earnhardt, for giving 'em what they come to see, every dang week."

Never has that sentiment rung more true than it did last year. By now, anyone with even a passing interest in NASCAR knows the story: a horrific crash at Talladega in July left Earnhardt to finish out the season with a broken sternum and a fractured left collarbone. Things did not go particularly well; in the 13 Winston Cup races after the wreck, he recorded only two top-five finishes. In many events, he was a decided, if uncharacteristic, non-factor.

The injury, Earnhardt says, was partly to blame, but *only* partly. Too often, he admits, he and his team just failed to get their car lapping fast enough. "I can't explain the bad run we had," he says. "I don't know what the hell got into us."

Still, he did some of the best driving of his life in the second half of 1996. His painful sprint to the pole position at Watkins Glen just two weeks after being hurt was nothing short of heroic. Of course, being Earnhardt, he one-upped himself on race day by leading the most laps, refusing to yield his seat to a relief driver, and finishing a

Nigel Kinrade

stunning sixth. Come autumn, he was brilliant at North Wilkesboro, where he lost a duel for the win with Jeff Gordon, and at Charlotte, where he qualified poorly but charged to the front and was a legitimate threat to win before fading to sixth. At year's end, he still managed to finish fourth in the Winston Cup standings, his sheer grit having salvaged the season.

The entire campaign mirrored the basic theme of his life, which has been this: you might be able to get Dale Earnhardt down, but *keeping* him down is another story.

"I think I'm at my best when the pressure's on," he says, a cockeyed grin spreading beneath his trademark mustache.

And the pressure has been on for over 25 years, from the days when he raced the short tracks in the shadow of his famous father, Ralph Earnhardt. When a heart attack killed Ralph in 1973, Dale was left to make it on his own. He managed . . . barely. His restless need to race cost him everything back then, including two marriages. But he was in a good Winston Cup ride by 1978, was NASCAR's Rookie of the Year in 1979, and went on to earn series titles in 1980, 1986, 1987, 1990, 1991, 1993, and 1994, tying Richard Petty as the sport's only seven-time champions.

He has earned more prize money than anyone in motorsports history, some $30 million and counting. He has banked a gazillion more dollars over the years from the sales of souvenirs bearing his name and likeness. He owns, among other things, a private jet and a couple of huge boats; a Chevrolet dealership; a 400-acre farm on which are raised Black Angus cattle, quarter horses, and Perdue chickens; and a race team fielding NASCAR Craftsman Truck Series entries for Ron Hornaday, Busch Series cars for Steve Park, and short-track rides for his three oldest children, Kerry, Kelley, and Dale Jr. His third wife, Teresa, and their eight-year-old daughter Taylor have him feeling "as happy as I've ever been." And yet he continues to race, and live, like a man desperate to make both a living and a name. In some ways, he is.

Giving 'em what they come to see: Earnhardt on the gas. *Nigel Kinrade*

Sitting in the cockpit of a race car, Earnhardt says, is "just natural for me." *Harold Hinson*

"I think you work better in debt, or when you're trying to achieve something," he says. "If you don't go out to achieve things, you'll never succeed. You'll never get any further than, 'OK, I've gotta go to work at eight, and I'm getting off at five. Ho-hum.'

"Me, I'm ambitious every morning I wake up. I don't have to get up until I want to, but I'm always awake at five or five-thirty, thinking, 'I've got a lot of things I want to get done today. I've got cattle, I've got farm things going on, the office is running wide open, we've got the race teams, the dealership.' I look at all that, and then I think, 'I've got to leave this afternoon to fly somewhere to practice and qualify tomorrow.' So I get up, and I go to work."

This ambition of his, this drive, must be at least partly responsible for his iron-clad connection with his fans, most of whom have to get up each day and go to work themselves. Last October, some 10,000 of them took a rare Monday off to attend the annual open house at Dale Earnhardt Chevrolet, and he ended up signing autographs for at least a third of them. It took him six hours and Lord knows how many felt-tip pens.

"We said we were gonna sit and sign for everybody who was in line, and we did," he says. Then he laughs, "I didn't realize it was gonna take 'til eight o'clock at night. But I've got a lot of good fans, the kind of fans who will stick with you whether you're hurt or whatever's going on."

Sensing an opening, I try out my theory on him, floating this notion that perhaps these people love him as much for his work ethic—*for giving 'em what they come to see, every dang week*—as for the things he has won, Earnhardt just shrugs.

"Maybe so," he says. "Maybe so. I don't know. I've never thought of it that way."

But if Earnhardt is reluctant to blow his own horn, those who know him best are quite happy to blow it for him. Car owner Richard Childress, with whom Dale has won six of those seven Winston

Cup titles, has been answering questions about his driver since they teamed up in 1984, and yet he still talks with a hushed awe about the tools that separate Earnhardt from the rest of the pack.

"He's definitely got a lot of ability," Childress says, "and you've got to have some natural ability to be a good driver. But you need to have the right attitude, mentally, to be a *great* driver, like Dale is. It's determination, is what it is."

Childress has formed the opinion that while the talent came naturally—"because he came from a racing family, and he got to see the approach that his daddy had to racing"—Earnhardt polished that talent "by going out and running every little short track in the country. He'd race every opportunity he got, and that's how you get to be good in this sport. Dale always had a lot of natural ability, but he tuned it over the years."

It was in the mid-1970s that Childress, then a struggling race driver himself, first encountered Earnhardt in one of Dale's first Winston Cup outings.

"I knew right then that he had everything it was gonna take to be somebody," Childress says. The vision stuck, and when he heard that Earnhardt was ready to bolt J. D. Stacy's team in the middle of the 1981 season, Childress jumped at the chance to hire him. He saw immediately that "we had the same goals and ideas. We both knew where we wanted to be someday."

But they were not yet capable of going there, at least not together. Childress' operation, then only a shadow of the powerhouse it has become, wasn't nearly as good as its new driver.

"I just wasn't ready to carry him," Childress says. "I knew that Dale had the ability to go on, and I wanted him to do that."

So Earnhardt took his helmet bag to Bud Moore's team, and spent two seasons there. In 1984 he rejoined Childress, who by now was fully fortified for battle. In the dozen years since then, they have been NASCAR's toughest team.

"We knew it was going to be a good combination," Childress chuckles, "but I don't think either one of us knew it would be *this* good."

And time has only improved the partnership, he says, because "Dale has gotten smarter on all the little things: the moves you make, when to go, and when not to go."

I ask Childress how he thinks Earnhardt has changed in the last dozen or so years, not as a driver but as a person. He kicks the question around, and then says, "If anything, he doesn't have the time today that he used to have. Being a champion, being Dale Earnhardt, he has to give up so much of his personal time. But you know, he's handled it pretty well for just being a small-town guy from North Carolina."

Which, most folks tend to forget, is exactly what Earnhardt is: a small-town guy who climbed up and out and became an honest-to-God living legend. It bothers Don Hawk that some people might see the latter at the expense of the former, that they will see the myth who, as Steinbeck had it, has emerged ahead of his accomplishments, and miss the flesh-and-blood human being beneath the fire suit and the helmet. It bothers him that there is always a built-in distance between a superstar and his fans, and that, viewed across this distance, the superstar can appear aloof, even cold.

"People see Dale at the track," Hawk says, "and he's got his game face on. He's intense, and he's focused."

But the game face, Hawk insists, masks an Earnhardt he describes as "loyal and generous. I've seen his loyalty demonstrated over and over again, and I've seen his generosity through the charitable things he does that nobody ever hears about. See, Dale is not one of those people who says, 'Here's my check, here's my check.' Sometimes I've happened to get a letter of thanks for something he did, and he'll tell me, 'Don't say anything to anybody about this. And throw that letter away.' Well, if I've seen some of these things, how many haven't I seen?"

Hawk, a deeply religious man, says, "There's a whole side of Earnhardt that a lot of people never see. And, unfortunately, he probably chooses to keep it that way. But I feel—and I don't use this expression lightly—I feel blessed by the Lord to be able to have seen glimpses of the real Dale Earnhardt."

With Teresa, Earnhardt claims to be "as happy as I've ever been." *Nigel Kinrade*

The real Dale Earnhardt, 45 years old and twice a grandfather, is still tough enough to kick your ass from Charlotte to Sears Point. But, let's face it, 45 is 45, and so 1996 ended with Earnhardt feeling what he called "healing pains" in his left shoulder, souvenirs from the Talladega tumble.

"I have a little bursitis in that shoulder anyway, in the outer joint," he tells me. "I crashed a street car, a 1956 Chevrolet, when I was 18, and bruised my shoulder up pretty bad. I've had just a touch of bursitis since then, and I think the [Talladega] wreck aggravated it."

I ask him about another of the crash's souvenirs, one of the most indelible memories of the entire 1996 Winston Cup season: the moment during the Brickyard 400 at Indianapolis, just six days after Talladega, when he climbed from the Childress Monte Carlo and allowed Mike Skinner to finish the race for him. I want to know if the whole thing was as emotional for him as it looked to the rest of us.

"It hurt my pride," he says softly. "I didn't want to get out."

He pauses. "I grew up proud. My dad was a very proud person. My family's always been pretty proud people. There's a sense of pride..."

It is something we have seen from him before. He broke a leg in 1982, never said a word about it, and kept on racing. He did the same thing when he fractured a vertebra in '95. And so when the Brickyard embarrassed him, he had no choice but to carry on, and to produce for us his startling qualifying run at Watkins Glen. Make no mistake, pride was as responsible for that pole position as a good chassis and a strong engine.

The crazy thing was, for all of the pain he was in as the season rolled along, only at Indianapolis did he seem to have a truly difficult time with the physical act of climbing into his race car.

He hasn't won since Atlanta in March, 1996. Earnhardt's belly is full of fire for 1997 as a result. *Nigel Kinrade*

From then on, he slipped right in, quick as you please, with his familiar one-step motion. I tell him that even when he is banged up, he looks more at ease crawling through that window than most healthy drivers do. He just smiles and says nothing, so I push the matter a little further, offering that perhaps this is simply what he was born to do.

Earnhardt tried to duck this, of course: "Aw, I've been climbing in and out of race cars ever since I was a young 'un. I was always climbing in and out of Daddy's cars, playing in 'em."

He is quiet for an instant, and I am, too. I want to see where he goes with this. Being Earnhardt, he doesn't go far.

"I don't know," he says finally. "It's just natural for me."

It is *all* just natural for him: the driving, the winning, the crashes, the injuries, the rebounds. It comes as easily to him as breathing does to the rest of us. And maybe that is why Earnhardt is able to wear this larger-than-life persona as casually as he does. He hasn't ever really figured out how to act like the legend he has become, because he didn't sit down and plan any of this.

"I've never been a daydreamer," he says. "Never even daydreamed about winning races. I've learned that you reap what you sow. When you work hard, you win. Daydreaming won't get you nowhere."

He talks some more about how he always does his best work under pressure, citing last October's Charlotte race as an example. He had opened the weekend by turning in a miserable qualifying lap, good for only 34th on the grid, and admits that he was feeling low as he drove out of the track that evening. When he got home to the farm, he noticed some activity at the race shop.

"Dale Jr. was working on his car, so I stopped in and talked to him a little bit," Earnhardt says. "I walked out of there and I saw that the lights were on in the shop behind my neighbor's house, so I drove over to see him. He was working on a '57 Chevrolet he's restoring. I sat around there for a half-hour, shot the breeze with him and his buddies. Then I went home, ate some supper with Teresa, talked a while, and went to bed."

"Earnhardt gives the fans their money's worth," says Don Hawk. "He may not win the race, but they know he's going to give them a show." *Nigel Kinrade*

By that point, he says, "I had gotten a lot off my mind about qualifying. I did it by doing the simple things that a lot of guys do: they get off work and they go home, or they go over to their buddy's place and hang out.

"By the next day, I was ready to go again."

I'm ambitious every morning I wake up . . . I'm always thinking, I've got a lot of things I want to get done . . .

That Sunday, Dale Earnhardt went out and did the only thing he knows how to do: he gave 'em what they came to see, driving clear to the front of the pack. It is something he plans to keep on doing for a few more years yet.

He frowns. "You know, somebody might look at the season we had in 1996 and say, 'Boy, Earnhardt's heading down the hill.' But we won races, and we were competitive. We just had a tough stretch after our accident. But we came out of it alive and well, and we'll go into 1997 even better."

There is no way to doubt him. No way at all.

16

STOCK REPORT

A Prognosis for Earnhardt—and Other Musings

BY BOB MYERS

Circle Track January 1998

The blackout that Dale Earnhardt experienced at Darlington typified the substandard year the No. 3 Childress Racing team endured to that juncture and beyond. The Ironman's brief and temporary physical impairment, and the once near-invincible team's slump were unexpected and are unexplained.

Most important, however, Earnhardt—after submitting to a battery of tests involving 25 doctors, 16 of whom he visited—was given a clean bill of health, bringing a collective sigh of relief to the driver and the motorsports world. The thoughts weighed heavily on Earnhardt for a few days that he might not drive again, and that his father had died of a heart attack at age 45.

No cause or abnormal effects were found from what was described by Dr. Charles Branch, a neurosurgeon, as a temporary dysfunction of Earnhardt's brain; a migraine-like episode involving a blood-vessel spasm or a short-circuit from an old injury. Such an unexplained neurological event is fairly common and may or may not happen again, the doctor says.

It may take more than 25 doctors to diagnose the ills of the team that was Winston Cup's benchmark for nearly a decade. Compared with lofty standards, the team blacked out too. The hope of Earnhardt's record eighth title, seventh for the Childress team, was buried in sixth place, 565 points deep, marking the first time they've gone three consecutive years without the crown. With a handful of races left, Earnhardt faced the prospects of ending the season with a career-longest, 59-race, nonwinning streak, and breaking his consecutive-year win streak at 15. It's not so much that the numbers are uncharacteristic—for example, four top-fives in 26 races—it was almost bizarre in several outings to watch the "Man in Black" as the man in back, struggling among the also-rans. So accustomed are we to seeing "the Intimidator" up front in the hunt that sometimes it was as if he wasn't at the race at all.

The "Man in Black" has struggled to not be the man in back in 1997. *Sam Sharpe*

Earnhardt is 46. History clearly shows that even the superstars inevitably reach a point when they win infrequently, or not at all. Maybe the reflexes have slowed a tad and, considering he's won 70 races, seven championships, and $30 million in prize money, maybe he doesn't take the risks of yore. As we age, life becomes more precious, no matter who we are or what we do. But I don't think Earnhardt has lost his edge as a driver, or his fire and hunger, and I wouldn't dare bet against his winning that eighth title. And Earnhardt's army remains strong and loyal. After finishing second from a 33rd start, with a boost from fuel mileage at Dover, Earnhardt detoured in front of the grandstands to a rousing standing ovation.

The woes of No. 3 seem to be a combination: turnover, a new crew chief, chemistry, and the competition, among them. Only a couple No. 3 crewmen are left from the most recent championship team in 1994, breaking up the pit and road crews. Larry McReynolds is a proven winner as crew chief, but he was not familiar with the chassis under the team's Chevrolets, or with the crew members, and he had not worked with Earnhardt. Admittedly, Earnhardt is a lousy qualifier, and McReynolds says that's a liability with track position being so vital, but he thinks it's the car, not the driver. Then, Jeff Gordon, heir to Earnhardt's throne, and his Ray Evernham-led Hendrick outfit, a cut above everything else, are everybody's problem. It's a mistake and foolish, though, I think, to write off Earnhardt and No. 3. Remember 1992, when Earnhardt plummeted to 12th in points with a sole victory and we asked, seriously, what was wrong with the team? Well, nothing that back-to-back championships didn't fix.

17

SIDEWAYS IN CHARLOTTE

Nap Time's Over

Circle Track February 1998

When Dale Earnhardt suffered what has been described as a lapse of consciousness at Darlington, he set off a storm of speculation and concern. Fortunately, the effects of the incident appear to have been minor and short-lived, but the furor that followed quickly developed a life of its own.

The sight of Earnhardt's famous black Monte Carlo chugging around the inside of the track while the rest of the competitors warily raced by was, to say the least, disconcerting to those in the stands and on pit road. Seeing Dale being hurriedly carried through the garage area to the infield care center did little to reassure anyone. During the first caution period, many of the drivers were more concerned about the condition of Dale Earnhardt than with the handling of their cars.

Earnhardt is regarded by most of his fellow competitors as one of the toughest people to ever sit down in a race car. When Dale was injured in a crash that many thought to be unsurviveable during the spring race at Talladega, some of the other drivers were reminded of their own mortality.

When Dale returned to race the next week with one of his best qualifying efforts in recent memory, many were visibly relieved. Ironhead was going to be OK. Then, when he again wound up on his head, this time at Daytona, only to jump out of the ambulance and get back in what could loosely be described as a race car, we all thought the old Ironhead was back.

Speculations throughout the 1997 season on what was wrong with Dale Earnhardt's team suddenly turned to what was wrong with Dale Earnhardt himself. Nodding out at Darlington indicated to some that perhaps Dale's on-track performance was the result of a lingering physical problem that culminated at Darlington. All this speculation was without foundation, and emphatically denied by all close to him.

However, is Earnhardt's performance really that bad? There are many in the garage area at every race who would love to have the finishing record Earnhardt has amassed in

Nigel Kinrade

1997. At this writing (after 24 races), Dale is in the top 10 in championship points with 11 top-10 finishes, and more than $1.2 million (an average of $51,490 per start) in winnings. In the most competitive series on our planet, I'd say Earnhardt is doing pretty well. Not what we are used to from the Intimidator, but good nonetheless.

For some reason, the fans and media keep pointing to Earnhardt's poor qualifying performance and wonder why. Earnhardt himself has said repeatedly that he is not a good qualifier, something that is not uncommon in racing. He has also stated that practice and testing are less interesting to him than racing, which may contribute to his qualifying woes. If there are no other cars on the track to race, he may simply not get pumped up enough. I watched qualifying at Charlotte Motor Speedway with Dale's brother a couple of years ago, and we decided that painting the fender of another car on his right-side window might be inspirational.

We also have to remember that there has been turmoil on the Goodwrench team during this season. They have lost at least two of their over-the-wall personnel, something that can be devastating to the best of teams. In addition, Larry McReynolds, aside from the problems in the crew, has had to try to figure out what one of the best seat-of-the-pants drivers in the sport needs in a car. That job may sound easy, but many believe the better the driver is, the harder it can be to get the car right.

If Earnhardt's recent performance is indicative of the future, his winless streak will be broken by the time you read this. Rumors hinting that Dale is growing tired of driving, and that his retirement is close at hand, gained velocity after he nodded out at Darlington. I think Dale will be around, and winning, in the foreseeable future. So far, all the Darlington incident has shown me is that maybe Dale just needed a nap.

18

DALE EARNHARDT

Stock Car Racing March 1998

The Seven-Time Winston Cup Champ Says He's Got a Lot of Winning Left in Him. You Want to Argue?

Let's be fair about this: If Dale Earnhardt is not always such a gracious loser, maybe it's because he hasn't really had much practice. From the time his career truly blossomed a dozen years ago, he has finished a Winston Cup season outside the top three in the final standings just three times. Only twice since joining the series full-time in 1979 has he gone through a season without a win.

When you get to Victory Lane so often, 70 times in all, maybe no place else feels comfortable. Least of all Manhattan when NASCAR is strutting its stuff for its big-city hosts, and all you want to do is forget your double-whammy season: fifth in points, no wins.

"By my standards," Earnhardt said, "it was a terrible year."

He was sitting on a sofa in his suite at the Waldorf-Astoria. It was a little after 10 o'clock in the morning. In about eight hours, he would make his way down to the hotel's grand ballroom for the annual Winston Cup awards banquet, the one evening when every driver is expected to be pleasant whether he is celebrating a great season or burying a bad one. I got the feeling that for Earnhardt, it was going to be a long night.

NASCAR has been holding these year-end dinners in New York City since 1981, and I have been to most of them in the course of one assignment or another. Never before had I come just to talk to the guy who finished fifth in points. But this was Earnhardt, a seven-time champion on the heels of a bum season that, in its own way, had created as loud a public buzz as Jeff Gordon's title run. Put two race fans together on Monday morning, at the office water cooler, or out next to the coffee truck, and before long they are talking about how Earnhardt ran yesterday. Winning or losing, he is always in the frame.

His team owner, Richard Childress, is often quizzed about Earnhardt's appeal. Childress responds, "Why did Elvis have the following he had? It's *style*." In Dale Earnhardt's case, style is a blend of brooding charisma, a willingness to race as hard or harder than the next guy, and a complete refusal to surrender.

Nigel Kinrade

A couple of years back, I asked him whether he thought there was a difference between drivers who went into a given race hoping to win, and drivers—like him, I figured—who were more concerned about *not losing*. His reply, short and not-so-sweet, became one of my all-time favorite Earnhardt lines.

"It's not in the *plan* to go out there and lose," he said.

Alas, as 1997 proved, the best-laid plans of mice and men and stock car legends often run straight into the wall. But now the season was over, and would be just a memory once this banquet was out of the way. All Earnhardt had to do was make it from the smoked salmon appetizer to the raspberry mousse, and listen to those last few nagging questions about where it had all gone wrong. Oh, yeah. Long night.

Just a few weeks earlier, in the days prior to the season finale at Atlanta, one quick moment on the TV screen told you all you needed to know about Earnhardt and 1997. He was standing with Jerry Punch outside the sprawling new headquarters of Dale Earnhardt, Inc., the corporate umbrella under which you find most of his holdings: the Chevy dealership, the 400-acre chicken and cattle farm, the rental properties, and the race teams, including the new Cup outfit for rookie driver, Steve Park.

Punch was taping an interview for ESPN2's "RPM2night" program. He asked Earnhardt about the enormous glass-faced building behind them, and then they discussed Park's stunning success aboard Earnhardt's Busch Series cars. And, because the piece was scheduled to air in the week leading into Atlanta, they chatted briefly about Jeff Gordon, Mark Martin, Dale Jarrett, and their battle for the Winston Cup title.

The session was clearly winding down when Punch, always the nice guy and never an ambush-style interviewer, said, "The last question I've got to ask you is, while all those guys are racing for the championship, can Dale Earnhardt . . ."

Earnhardt, smirking, cut him off. "I don't want to hear it. *'Can Dale Earnhardt do it?'*"

There was nothing rude about his interruption; he and Punch are old friends. It was just that Earnhardt had seen the question coming. It had been popping up, after all, since his last win in March of 1996, and particularly since a Talladega wreck in July of that year seemed to trigger a slump from which he had never recovered. But even as Earnhardt chuckled about having correctly guessed the question, there was no mistaking his disdain for the topic.

"Yeah," he said to Punch, and to the world, "he can do it."

As the record shows, he could not. Earnhardt ran awfully well at Atlanta—his early charge to the front was spellbinding—but he finished 16th after glancing off the fence. And so the questions went on.

Now, in his suite, I asked, "How tough is it when all anybody wants to talk about is this slump?"

Nigel Kinrade

"It's the nature of the beast," he said. "The nature of the press."

Earnhardt sighed. "I don't like to analyze things. When we have a bad race, I just say, 'OK, this is what happened. Let's try to fix it.' And I go on from there."

Which is a fine philosophy to live by, but not the most realistic expectation when you are the biggest name in a booming sport. Whether he likes it or not, Earnhardt's seven championships—in 1980, 1986, 1987, 1990, 1991, 1993, and 1994—make his current slump a very real issue in the NASCAR garage area, and in homes across Motorsports America. And so the press, the beast, has little choice but to push him toward the kind of analysis he hates.

Most often, Earnhardt will play along until he is tired of the process; then he simply stops giving answers and delivers non-answers instead. The events of last autumn are a perfect example: on the opening lap of the Southern 500 at Darlington, Earnhardt suffered some sort of a blackout that doctors eventually dismissed as a "migraine-like event" not likely to be repeated. Cleared to drive in the next race at Richmond, he strode into the media center and answered, as best he could, every medical question thrown at him. Addressing it up front, he figured, would end the matter once and for all. He was wrong. But the press was wrong, too, in believing that it could badger more information out of Dale Earnhardt. As the weeks rolled on, he began to joke that he had been "beamed somewhere" by aliens who had snatched him as he slept on the eve of the Darlington race. Before long, the beat reporters understood that they had

When Larry McReynolds was introduced as Earnhardt's new crew chief last winter, he was thought to be the missing piece in the championship puzzle. It didn't work out quite that way. *Nigel Kinrade*

"The teams that have run for the championship, they're contenders to win every race," says Earnhardt, here battling Jeff Gordon. "That's how you've got to be, and we were not." *Nigel Kinrade*

gotten all they were going to get out of him about the entire affair. It was typical Earnhardt: victory by non-answer.

On this Friday morning in Manhattan, it happened in reverse, with the non-answer coming first. I had thrown him a question about 1997, and Earnhardt's initial reply was a shrug, followed by this: "It just didn't go the way we thought it would. There's a lot of reasons, but I don't think there's any use in getting into 'em. To me, it's water under the bridge."

But then, after just the slightest hesitation, he waded right in.

"I think what happened is that in 1993 and 1994, our team was running along on a certain plane, a smooth plane. Things were going good. Then, boom, along came 1995."

That was the season in which Gordon and his Rick Hendrick team reached full strength, winning seven races to Earnhardt's five and earning young Gordon his first championship; Earnhardt was the runner-up. In 1996, Terry Labonte gave Hendrick another title, edging teammate Gordon; Earnhardt, in the thick of the points chase up until Talladega, slipped to fourth.

"The competition level got stepped up a notch [by the Hendrick camp], but our team stayed on that same plane. We never changed. As we went through '95 and '96, we got behind in a lot of areas."

The Talladega crash, which busted his collarbone and fractured his sternum, would make a dandy excuse if Earnhardt wanted one, but he does not.

"People point to that wreck and say, 'He ain't driven the same since then.' Well, I bounced right back and won the pole at Watkins Glen, and we were competitive at other places. But we were not staying on top of things. Our team got off track, and as we got off track, it frustrated me. We just sort of got confused a while."

Earnhardt said, "Look at the teams that have run for the championship: the 24 [Gordon], the 5 [Labonte], the 6 [Martin], the 88 [Jarrett] this past season. They're contenders to win every race. That's how you've got to be, and we were not that way in '95 and '96. So in '97, we went to work on getting things going in a new direction, and Larry McReynolds was part of that direction."

McReynolds, hired away from Robert Yates' Ford team in November of 1996, was thought to be the crew chief who could get Earnhardt and Childress back in the groove. For most of one Sunday afternoon last February, the move looked brilliant: "The Daytona 500," Earnhardt said, "was gonna be a good race for us." Then he slid into the wall while fighting Gordon and Jarrett for the lead, and flipped down the backstretch.

His fans cheered the way he soldiered on to finish that race, his wrinkled Monte Carlo held together mostly by black duct tape. What few of them realized was that Daytona would be a metaphor for their man's entire season: flashes of brilliance here, bitter disappointments there, the only constant being Earnhardt's unceasing drive to make the most of what a race day handed him.

He said, "It took longer than we figured it would for Larry to get a grasp on things: chassis setups for me, the aerodynamics of the Chevrolets. It frustrated me that he and I didn't get on top of things as quick as I thought we would. But Larry is a brilliant person, and there's nobody as dedicated as he is to getting the job done. I still have all the confidence in the world that he and I can do this.

"Don't forget, Richard Childress and I have showed we can bounce back. Go back and look at it: in 1986 and 1987, we won the championship. In 1988 and 1989, [Bill] Elliott and Rusty [Wallace] won, but we won it again in 1990 and 1991. In 1992, Alan [Kulwicki] won it and we were in a bad, bad slump, but we came right back and won championships in 1993 and 1994.

"So, yeah, 1997 was tough, but we can bounce back. And we will."

Of course, Earnhardt is hardly the first stock car hero to talk about rebounding from hard times. Richard Petty did it all the time in the late 1980s, and Darrell Waltrip has spent much of the mid-1990s threatening to win another race. But the difference between those two and Earnhardt is both obvious and enormous: it pains Earnhardt to even acknowledge his age; Petty and Waltrip have talked so freely about their golden years that I often wondered if it didn't contribute to how poorly they were doing at the moment.

I have this theory that as soon as somebody hangs the tag "elder statesman" on an athlete, the guy may as well head for the showers. Call it coincidence if you want to, but the winning stops. I ran this past Earnhardt, without naming any names, and he nodded.

"They talk themselves into it," he said. "Or maybe they let other people talk 'em into it."

They buy into the elder statesman rap. And the next thing you know, they are struggling to run in the top 20 during their own farewell tours, as Petty did in 1992; or they are complaining, as Waltrip did in October at Charlotte, that somebody else took the last available provisional starting spot.

I know this: those were not grand moments to tack on at the end of dignified careers.

And I know this: the first guy who calls Earnhardt an elder statesman had better duck. He will turn 47 this April, and yet he still has more of an edge to him than any rookie in the garage area.

Generally speaking, Earnhardt is a pretty decent interview, at least until the questions become routine. *Nigel Kinrade*

I asked him if he believed that stuff about being as young as you feel. Earnhardt said, "Well, I've never felt old. And I don't just mean in racing, I mean in life. I can relate to the young workers, the young racers, the young people around me."

He laughed. "I'm not an old fogey."

Before Earnhardt, the last great champion who refused to talk himself into a rocking chair was Bobby Allison. He was forced to retire in 1988 at age 50, but only after a Pocono crash damn near killed him. What is important here is that right up until the moment of that wreck, Allison was as fierce and stubborn a racer as he had been as a first-time Winston Cup winner in 1966. A man who endured his share of slumps, Allison is amused to hear people making so much of the one Earnhardt has been in.

"I can sympathize with him," Allison says, "because there were key people who wrote me off whenever I had a bad season, even when I was in my 30s. But I won the Winston Cup championship when I was in my 40s, won the Daytona 500 when I was 50. And I still wanted to race, and race hard.

"You know, Earnhardt used to be the second most committed guy in racing. Then the most committed guy got hurt. And since then, no one has shown the commitment or the intensity that Dale has."

Earnhardt said, "If you go just by time, I'm one of the oldest drivers on this circuit. I know that. But I ain't done yet. I feel as competitive as I always have. I feel as determined as I always have. I don't give out during the race. I don't get out of the car tired and worn-out. I'm always eager to race some more."

I said, "We've seen drivers at similar stages in their lives enter into slumps they never rebounded from. Have you had any moments when you've said, 'My God, is this *the* slump?'"

"You mean, like, 'Is this the last one?' Nah, I mean, you might *think* about it, but I don't *feel* it."

We talked a bit about Allison, about his ups and downs. At one point I said, "The thing about Bobby is, he never acted like a past champion closing out his career, never played that role."

"And I'm not ready to play it, either, " Earnhardt snapped.

You have to wonder if he ever will be. Sitting there, I thought about Richard Childress and his Earnhardt-as-Elvis line, and how it fits only until you think about what eventually happened to Presley: he got old, got fat, and no longer sang with the energy that had helped him burst into prominence. He let himself be crowned an elder statesman of rock 'n' roll, and that was that.

Maybe I am wrong, but I do not see any of this happening with Earnhardt. Not this year, not next year, not ever. I do not see him talking his way into that rocking chair. I do not see him hanging on too long, scraping his way into races, finishing four or five laps behind when he finishes at all. I do not see him claiming against all evidence, as so many elder statesmen before him have, that he might just have a win left in him yet. I do not see a grand farewell tour, however tempting the souvenir sales might be.

I filed away something Earnhardt said last October, when basketball coaching legend Dean Smith retired from his post at the University of North Carolina with a simple announcement.

"Ol' Dean was pretty cool, wasn't he?" Earnhardt remarked at the time. "He said, 'Well, it's time to quit now.' I think when I quit, I'll just quit."

That will not happen, he stressed, before his present contract with Childress runs out after the year 2000. But if he looks at things at that point and sees the candle flickering, even just a little, I hope he blows it out altogether. To do otherwise would tarnish one of the last honest images in sports.

But I was getting way ahead of myself. Because here was Earnhardt, fitting that image perfectly, leaning forward, his voice a hiss: "There ain't just one or two wins left in me. There's a *lot* left."

That is also the opinion of one Bobby Allison, who says with a been-there done-that wink, "I don't think we should go writing off Dale Earnhardt just yet."

Let us hope not, because it is jarring to imagine Winston Cup racing without Earnhardt in it. He still throws the best moves in the game; in that 1997 closer at Atlanta, his battle with Ward Burton had the entire sport on the edge of its collective seat.

Waltrip says of Earnhardt, "He can do more with a car than anybody I've ever raced against."

And he is still a necessary ingredient in every significant rivalry, real or hyped, that you can name: Earnhardt and Waltrip; Earnhardt and Wallace; Earnhardt and Irvan; Earnhardt and Martin; Earnhardt and Bodine; Earnhardt and Elliott; and Earnhardt and Gordon.

All he has been, across the last dozen years, anyway, is the biggest star in the business. How big? His agent, Don Hawk, claims that "ESPN, TNN, TBS and a bunch of other [television] companies have come to us saying, 'Let us do the life of Earnhardt.'"

I have written this before, and I write it again here: drive down any street in any town in this nation, and before long you will spot a Dale Earnhardt bumper sticker. For a self-described country boy from Kannapolis, North Carolina, he has left a pretty significant footprint on the landscape.

But more than that, we need him around because Earnhardt is the last real cowboy left in NASCAR, the last guy willing to play almost exclusively by his own code.

For much of the late 1980s, while he was emerging as something more than just another good driver and his take-no-prisoners driving style was transforming stock car racing, NASCAR made serious attempts to smooth his rough edges; in one period, its officials seemed determined to slap his wrist every time he creased a bumper. But if any of this tough love had an impact on Earnhardt, it didn't show. He kept the hammer down, kept trading paint, and kept winning. No one, least of all some guy behind a desk, was going to tell him how to drive a race car.

And no one, least of all some guy behind a microphone or a notebook, is ever going to tell him how to think. This is part of Earnhardt's curious, arm's-length marriage with the media; he is fairly accessible, considering his stature and his schedule, but he has no patience for questions that are routine or, worse, leading.

On a handful of occasions in 1997—at Daytona, Charlotte, Dover, Talladega, New Hampshire, Martinsville—Earnhardt looked like a potential winner all day long. Packs of reporters descended on him each time, hoping to prod him into captivating quotes about how close he had come to breaking his dry spell. They might as well have asked him to sing a Broadway show tune.

"I don't play what-ifs," he said at the Waldorf. "Never have."

No other driver—at least no other superstar driver—is so independent about such things, and this makes Earnhardt a vital late-1990s commodity. You hear a lot of talk these days about team owners searching for the next Jeff Gordon, meaning the next driver who is young and handsome and fast, and one of these days they are going to find him. But we may never see another Dale Earnhardt, may never see a guy whose code is so inflexible.

At this same hotel in 1993, Earnhardt was enjoying the sixth of his seven championship dinners when Ken Schrader walked to the stage to collect his rewards as a top-10 finisher in the Winston Cup standings. The two of them had some recent history: at Phoenix, the next-to-last race of the season, Schrader came out on the losing end when they tangled. Now Schrader, ever mischievous, stood at the podium and announced that he had several people to thank. He looked straight at Earnhardt, who was sitting at the head table, and said, "Special thanks to you, Dale, for helping me finish ninth. I thought I was going to finish *seventh*."

A lot of drivers I know would have shifted in their seats, unsure of how to react. The more PR-conscious among them would have broken into nervous tics.

Earnhardt? He just smiled.

"It's kinda funny," he said in Manhattan this time around. "In December, you're worn down. You're beat from the season and all the things you have to do all year long. And yet I'm eager to get started again. I can't *wait* to get to Daytona."

I said, "More eager than usual?"

"Yeah, probably," Earnhardt replied. "It's time to *win*."

You do not spend any amount of time around him without believing that he will come back strong. Stronger than ever? Probably not, because "stronger than ever" means a whole bunch when you are talking about 70 wins and seven championships. But to bet against him coming back is to bet against history, not to mention all of that commitment and intensity.

Dale Earnhardt has nothing left to prove—unless you count yourself among the nitwits who will consider his portfolio incomplete until he wins the Daytona 500—and, in real-world terms, very little to gain. He has banked over $30 million in race winnings, and that figure is dwarfed by his slice of various licensing and merchandising deals. According to *Forbes*, Earnhardt earned $19.1 million in 1997, placing him eight on the magazine's "Super 40" list of the world's highest-paid athletes; some $15.5 million of that was for endorsements, a category in which he ranked fourth behind Michael Jordan, Tiger Woods, and Arnold Palmer.

But none of that means a thing on Sunday afternoon, and Earnhardt has been spanked on too many Sundays lately. For that reason, and that reason alone, he will be worth watching in the early rounds of 1998.

I asked him, "Do you feel like you've lost anything at all as a race driver?"

He narrowed up his eyes, pondered the question a moment. Beneath his trademark moustache appeared another Earnhardt trademark: that wonderful cockeyed grin.

"The only thing these guys have got on me," he said, "is age."

That evening at the banquet, he accepted his point fund check with a quick, subdued speech. I went looking for him later, just to say goodbye, but I had no luck. Elvis had left the building.

"There ain't just one or two wins left in me. There's a *lot* left." *Nigel Kinrade*

19

THE 40TH DAYTONA 500

BY BENNY PHILLIPS
Stock Car Racing May 1998

After Trying for 19 Years, Finally Earnhardt Won Stock Car Racing's Biggest Prize

Dale Earnhardt's Daytona 500 win was one of NASCAR's grandest moments, and all of racing breathed a sigh of relief when it ended. Never has such drama unfolded on the high banks, and few events have matched this finish for excitement.

Most fans, and almost everyone else connected with racing, figured Earnhardt would win the Daytona 500 if he kept banging on the door. He won the race on his 20th try, and he did it in Earnhardt fashion, which was important to him. He led 107 of 200 laps and held off one competitor after another, who moved up to take a shot at the lead. But Earnhardt held the hill, like an honorable soldier in a pillbox. He led five times, including the last 61 laps.

The win was more than a milestone for the driver. As Earnhardt crossed the finish line, ending a 59-race winless streak that stretched back to March 10, 1996, an entire sport came to attention. And as he stood in Victory Lane bringing back all those memories of people and days gone by, he stood the entire sport on its top.

So, Earnhardt is back, and NASCAR gets all that comes with the show.

There were more story angles than laps in the race, and the drama cup overflowed.

In a story line similar to those home runs that Babe Ruth used to hit on request, Earnhardt accepted a good-luck penny from a sick child less than 24 hours before the 500. He carried the penny in his car on Sunday. "On Saturday they had five Make-A-Wish kids in the NASCAR office and we went to visit," Earnhardt said. "One little girl in a wheelchair—she was tiny but had a pretty voice—gave me a penny and said, 'I rubbed this penny and it's going to win you the Daytona 500. It's your race.'

"I glued that penny on the dashboard and it's still on the dashboard," Earnhardt said. "I'd like to take that penny with me to the rest of the races, but it's glued there and Lesa France Kennedy [who heads Daytona USA] says I can't have it back.

Daytona 500 winner Dale Earnhardt with the Harley J. Earle trophy. *Nigel Kinrade*

"All race fans are special." Earnhardt continued, "but a little girl who's in a wheelchair that life has not been good to, giving you a penny and wishing you luck, that's pretty special. Pretty special indeed."

The theater of victory touched many, including Earnhardt. "It took 20 years, and this is the greatest win of them all," said the man who had just captured the 71st Winston Cup event of his career. "I won't say I cried on the final lap, but my eyes sure watered up."

It was an emotional experience for Earnhardt when he came down pit road, too. Crew members stood in single file in front of their respective pit stalls to shake his hand as he slowly drove by on his way to victory circle.

"That was unbelievable," he said, "all those crews coming out to meet me."

Before he drove into victory circle, he cut right and did a couple of doughnuts on the perfectly manicured grass between the race track and pit road. NASCAR president Billy France, who was among the first to congratulate Earnhardt via the team's two-way radio, heard of Earnhardt's plan and told him, "That's fine with us."

France knew the value of this victory.

"I don't see how central casting in Hollywood could have done any better," France said. "Why is it special that he won it? Number one, he's won almost everything else but the 500. Number two, and perhaps more importantly, he's been so close so many times. He's been snakebit at Daytona. And topping it off, he was coming off a 59-race [non-winning] streak.

Jubilation was the order of the day in Winners Circle. *Nigel Kinrade*

"I think it was successful in just about every respect," France said. "The competition that we had all day, the lack of yellows, the high average speed, and the fact that we went through the whole 10 days without anybody going to the hospital except for a check or so—and nobody was there when Sunday night came—and then Dale winning the race. And we were lucky on the weather. So in all respects, we can't wait until next year."

On Sunday afternoon while Earnhardt was holding his press conference, somebody noticed that fans were digging up the green grass where he spun around. "We'll have those guys arrested," the speedway public relations director stated. "No, no," Earnhardt shouted. "I love it. I love it. Let those people alone. I'll pay for the grass if necessary."

Earnhardt averages 172.712 miles per hour. Three caution periods slowed the event for nine laps. The yellow lights blinked for the third time as Earnhardt led the parade down the backstretch headed for the white flag, indicating one lap to go.

"I was looking in my mirror and saw a car spin off the track in turn two," he said. "I knew the race would be over when we got back to the line. When I took the caution flag and white flag at the same time, I knew nobody could pass me, and that I was going to win if my car would make it back to the line. I slowed down in turns one and two and then speeded up again. I thought, 'Heck, let me get back around there to the finish.'"

John Andretti and Lake Speed had spun off turn two to bring out the final yellow. Earnhardt was leading at the time of the previous caution when Andretti and Robert Pressley tangled on lap 174. The field pitted for the final time, and Earnhardt came out first with teammate Mike

Last year, Earnhardt's crew performed miserably on pit road. This year, with new players and only gasman Chocolate Myers a longtime veteran, they were splendid. This was the last and most important pit stop. Earnhardt beat everyone out to retain the lead. *Staff photo*

After receiving congratulations from crewmembers, Earnhardt cut doughnuts in Bill France's lawn. France said he didn't mind. *Staff photo*

Skinner on his bumper. The cars lined up with Jeremy Mayfield third, Rusty Wallace fourth, Jeff Gordon fifth, Bobby Labonte sixth, and Jimmy Spencer seventh. There were 23 laps to go at the restart.

Teammates Mayfield and Wallace slipped by Skinner and tried to work over Earnhardt. Gordon, who led 56 laps earlier, became a player in the wild shuffle. He drove to the bottom of the track in an attempt to challenge for position. Wallace drove low, too, and bumped Gordon to the grass on the backstretch.

With 10 laps left, it was Earnhardt, Mayfield, Wallace, Labonte, and Gordon. Labonte and Gordon hooked up and tried to move, but Mayfield blocked the path. The cars scattered and Gordon moved up to second, but Mayfield charged back around.

Five laps to go and the lead pack consisted of Earnhardt, Mayfield, Gordon, Wallace, and Labonte. Gordon's car developed an engine problem, and he faded back to a 16th-place finish.

In the next shuffle, Labonte pulled up to second, passing Wallace and Mayfield. As the cars came off the fourth turn headed to the yellow and white flags, Mayfield ducked to the inside of Labonte. The two beat on each other for position all the way through the tri-oval with Labonte ahead at the line, finishing second. Mayfield was third, Ken Schrader (racing with a sternum broken during Thursday's 125-mile qualifiers) fourth, and Rusty Wallace fifth. Ernie Irvan, Chad Little, Mike Skinner, Michael Waltrip, and Bill Elliott completed the top 10.

"That played into my hands," Earnhardt said. "They got to racing each other, and I was able to get a little distance on them. Bobby maybe could have passed me, but he was pretty much out there by himself, and I felt I could hold off any of them one on one, but not four on one. We were fortunate they did get to racing, and Bobby was fortunate to come back to second. They were racing hard back there, and I was watching it all. I drove looking in my mirror most of the time. You can do that when you're in front. I had to drive with my eyes in the mirror, because there was somebody always trying to pass me. They kept coming and kept coming."

Labonte wanted another shot at Earnhardt. He may always wonder what might have happened had the 500 gone green another lap. As in Bob Dylan's song, "The answer, my friend, is blowing in the wind."

"We had a good run going," Labonte said. "I got to Earnhardt, and he was running about the same speed. I have to congratulate him. I had a feeling all winter he was going to win the 500. I guess I should have bet on him."

Earnhardt said he had the perfect car and perfect engine. "I was working to stay in front until somebody turned me over or I won the race," he said.

"I am happy, and I want to thank everybody who has had anything to do with my career. I especially want to thank my family, team owner Richard Childress, and the entire crew. We worked so hard to win this race. Now we're going for that eighth championship."

Earnhard was running out of thank-yous by the end of the day. "Skinner helped me on that last restart," he said. "I said my prayers today, and I thanked Him after the race. I think I thanked everybody who ever touched my life in racing."

Childress talked about the last pit stop: "When we came out in front after the stops, I knew Dale was going to lead the race all the way. I felt good when Skinner was behind him because I knew that would get us off to a good restart. But I got to looking at those two Penske cars, and I knew that wasn't good. Finally, they got shuffled apart, and I liked that."

Childress lost half a dozen employees during the winter break, and there are those who figured this would be the worst year yet for the operation. Crew chief Larry McReynolds was not among that group. "We pulled with one rope, and won the 500," he said. "We are a team that is pulling together."

Earnhardt qualified on the second row a week before the 500, on a day the Labonte brothers won front-row starting positions. Then he won his 125-mile qualifying race (Sterling Marlin won the other) Thursday prior to the 500. So what, the crowd said to that. Earnhardt has won nine straight qualifying races.

The most frightening moment in the event came late in the going when several cars tangled on pit road. Fortunately, no crewmen were hurt. Dale Jarrett's car and that of Jeff Burton were damaged and no longer competitive after this. *Harold Hinson*

Still, he said this year's 125 was a lucky one for him. "I was riding along in the lead thinking who would hook up and draft by me. Nobody was able to and I drove to victory. I can't explain how I have won nine in a row of these things and 11 altogether, and can't win the 500."

Earnhardt had been close so many times that you had to figure somebody out there in the world of black magic had whipped up one incredibly powerful potion with his name on it. Wing of a bat, eye of an owl, tongue of a lizard, three bear paws, and all that. But it only worked for the 500 because Earnhardt had already won every other major Daytona race.

You had to envision a cackling, stringy-haired old dame somewhere deep in the darkest swamp chanting over a bubbling cast-iron pot, using her long, bony fingers to hold up an Earnhardt voodoo doll. For 19 years, she pulled her pin from the pot the third week of February and stabbed the doll in the heart, making strange things happen to Earnhardt. It also made his team jump and twist and flinch with agony.

A half-dozen times since Earnhardt showed up as a rookie in 1979, he had the Daytona 500 checkered flag between the crosshairs. A couple of times he had dominated, his competition helpless to do anything but watch him win.

Then he would melt like an icicle in August.

He ran out of gas with a couple of laps to go in 1986, and in 1990 he was leading the race coming down the backstretch on the last lap and ran over a piece of metal from someone else's car. It cut a tire and he finished fifth.

The old dame continued to brew her concoction, adding a muskrat tail now and then. She was depending on Earnhardt's label as a driver who could not win the Daytona 500 to be printed permanently into the lining of his résumé.

Was it a curse?

"No, but that's a good question," Earnhardt said before the 500. "We work like the devil before this race, just like we work like the devil before every race. All you can do is look in the mirror and ask if you have done all you can do. If the answer is yes, then I rest easy."

So what made the 20th year different?

El Niño, of course.

The swamp was flooded and the old dame couldn't get to her black pot. And that, my friends, is as good as any answer you're going to get as to why Earnhardt won this year after so many years of being so close.

Ken Schrader proved he was the toughest guy in town when he ran the entire Daytona 500, finishing fourth, with a cracked sternum. *Harold Hinson*

Earnhardt was intense and focused all week. *Harold Hinson*

20

CAN EARNHARDT REPEAT

What to Watch for in the Great American Race

BY BRUCE MARTIN

Circle Track February 1999

It took 20 years for Dale Earnhardt to achieve his quest of winning the Daytona 500. When Earnhardt finally accomplished his goal last February, it stood out as one of the greatest moments in NASCAR's 50th anniversary celebration.

No longer could Earnhardt be asked the question, "Are you ever going to win the Daytona 500?" Or, "Will your career be incomplete without a victory in the Daytona 500?"

"This is my 20th year, and I'm tired of answering that question of why I haven't won the Daytona 500," Earnhardt said before last year's race.

Now that Earnhardt has finally put those questions to rest, a new question may be asked prior to the 1999 Daytona 500: "Dale, can you win the Daytona 500 two years in a row?"

Can Earnhardt repeat this scene in 1999?
Sam Sharpe

In NASCAR Winston Cup racing, success is ephemeral. For every accomplishment, another goal looms ahead, giving Winston Cup drivers or teams just a short time to experience their feat.

To say Earnhardt finally knows how to win the Daytona 500 would be inaccurate. Earnhardt entered last year's Daytona 500 as the all-time victory leader at Daytona International Speedway with 30 wins. In fact, Earnhardt probably knew how to win the Daytona 500 much earlier in his career, but a variety of circumstances, and just plain bad luck, kept him out of Victory Lane until last year.

Nigel Kinrade

Back-to-back Daytona 500 victories for Earnhardt?

If he were to be successful in winning Daytona two years in a row, he would join a select group that includes Richard Petty (1973–1974), Cale Yarborough (1983–1984), and Sterling Marlin (1994–1995).

"It does become easier to a point when you go back somewhere after you have already won that race," Earnhardt admitted. "I've won more races at that race track than I have anywhere else. I've won more races there than anyone else has. To go back there and win it after winning the Daytona 500, the one that had eluded me for 19 years, I do feel better about it. I am more excited about it. I don't think it gives me any advantage or edge. Everybody is as tuned at Daytona as I am. I think the testing and things that have gone on and are happening are very important.

"You look at the Richard Childress team and myself—we go out and win races. Now, we have not won races in a while, and last year didn't go the way it should have or we wanted it to. No, it's not easy to go out and do it again. You have to work hard every time to do it again.

"You don't win a championship again because you won it last year; you win it again because you work hard to win it again. The same goes for races. Daytona will be just as hard to win as it has been before."

There are many keys to success in the Daytona 500. It's a speedway that requires an outstanding restrictor-plate engine—a motor that can produce horsepower while being choked for air because of the smaller openings in the restrictor plate, which fits above the carburetor. The car must also slice through the air aerodynamically. Also, it must be good in the draft, which brings another variable into the equation—the value of a drafting partner.

"I have Mike Skinner," Earnhardt said of his teammate at Richard Childress Racing. "I think that at any aerodynamic track you run on, it's important to have a drafting partner or someone to work with in the draft. It's become something that you see the Ford teams, the Chevrolet teams,

A drafting partner is essential at Daytona. Dale Earnhardt got plenty of help from teammate Mike Skinner last year. *Nigel Kinrade*

the Roush teams, the Hendrick team, and the Penske team working on together. I think if we didn't have Mike Skinner in February, I wouldn't have won the Daytona 500, or my opportunity wouldn't have been as good.

"Mike got me out there, got me out front, and got me going. I think it was a major, major plus."

The art of a drafting partner also worked wonders for Jeremy Mayfield, who had his highest finish at the Daytona 500 with a third place last year while his Penske teammate—Rusty Wallace—also scored a career-high finish at the Daytona 500 with fifth place.

"Rusty and I helped each other a lot at Daytona [in the 1997 Daytona 500]," Mayfield recalled. "We were able to push each other and to help each other out. Sometimes I'd wait up to help him when he was in a little trouble; sometimes he'd wait up to help me when I was in a little trouble. He might make a couple of moves that would help me out, and I might make a couple of moves that would help him out. Our goal to the end was to be sitting one-two at the white flag. It didn't work out that way, but that's what we were shooting for. I don't know who would have won if it had worked out that way, but I do know there wouldn't have been a wreck, and there wouldn't have been any rubbing, and we would have finished one-two, too.

"Everything you do at Daytona and Talladega is with other cars on your mind. You're looking to your right, to your left, in the mirror, and through the windshield. You do all of these things to keep guys from doing things you don't want them to do. In the Daytona 500, Rusty and I were able

DALE VS. DALE (JR.)

By David Miller

As the 1999 race season kicks off at Daytona, it's not too early to look ahead at some things that forecast new and exciting race scenarios. Among such developments is the entry of Dale Earnhardt Jr. into the ranks of the Winston Cup competitors. To preserve his Winston Cup rookie status, Dale Jr. will be participating in only five Winston Cup races in 1999. Upon entering the top level of circle-track racing, one of the first things he will encounter on the track is his father—The Intimidator. Not since 1992 (Richard and Kyle Petty) have we seen father-and-son competition in Cup racing, so we turned to Dale Jr. to get his thoughts on racing with his dad.

"Racing in five Winston Cup races next year will be a good opportunity to get some experience in a Winston Cup car," he said. "I'm not really sure how much racing I will be doing against him yet. There will be a lot of other cars out there, but if I get up beside him somewhere it will be a lot of fun.

to do those same things so we would do what we wanted each other to do.

"I think the '97 Daytona 500 was a prime example of the way two teammates should run a race. I think Rusty and I are fortunate that most other teams don't look at it the same way. We came out way ahead because of the way we worked together. Believe me, it's nice having a friend out there."

It's obvious to see, whether in the grandstands or watching on TV, how the drivers in the Daytona 500 are constantly jockeying for drafting position. Through long stages of a race, drivers may be content to stay in the long drafting line because, without a drafting partner, an attempted pass can backfire and send a car from fourth to 15th place.

Men who are fierce rivals often become other race drivers' long-lost pals if they believe the two cars working together can improve their positions in the field against their rivals.

"You have to have a drafting partner at Daytona and Talladega," Mayfield explained. "You might go to the dance alone at those places, but you aren't going to be there very long if you do. Drafting is everything there. Having just one partner you can really trust doesn't give you an advantage; it makes you even with just about everybody else.

"Restrictor plates make the draft vital to doing anything at those tracks. You sure aren't going to drive off and leave anybody there. Two guys aren't going to run off and leave anybody there. Shoot, even three or four guys aren't going to run off and leave everybody there. At the best, a partner can help you and help himself move toward the front, but it still boils down to a game of chess at 200 miles per hour. What you are trying to do is not win the race with partnering and positioning. What you are trying to do is get yourself where you want to be for that final lap. It's kind of crazy. You spend 187 laps at Talladega trying to get yourself exactly where you want to be for the white-flag lap."

Starting position is very important at the Daytona 500, but not as essential as it would be at a

The GM Goodwrench Service Plus crew is unanimous on Earnhardt's chances of reaching Victory Lane again at Daytona. *Sam Sharpe*

track where it is very difficult to pass. The starting field in the Daytona 500 also is determined by the most complicated qualifying procedure in all of racing.

Only the front row is determined on Pole Day, which is held eight days before the Daytona 500. The remaining three days of time trials determine the starting lineups for the two Twin 125-Mile Qualifying Races held the Thursday before the Daytona 500.

The top 14 cars from each of the Twin 125s—exclusive of the front row set from Pole Day—form positions 3-30 in the Daytona 500 starting lineup. The finishing order from the first race lines up on the inside rows, while the finishing order from the second 125-miler takes the outside. Positions 31-36 are determined on the three days of time trial speeds, with starting spots 37-42 filled on provisional starting positions from the previous season's NASCAR Winston Cup standings.

The 43rd, and final, starting position is reserved for a former NASCAR Winston Cup champion who failed to make the field based on the previous criteria. It only sounds complicated because it is.

After Earnhardt won the Daytona 500 for the first time in his career last February, he believed he was on his way to a record-breaking eighth NASCAR Winston Cup championship. But it didn't happen, as 1998 was a season of frustration for Earnhardt and his team.

"This season has been rocky with highs and lows," Earnhardt said toward the conclusion of the 1998 Winston Cup campaign. "When you pull into a race track and they drop that green flag, hey, you want to be one of those guys the fans, the crews, and the teams are thinking, 'I've got to watch out for that No. 3 car.' Right now, we aren't that force in Winston Cup racing. We have to get back to that position.

"I don't know what happened to the team. You're going to have to go talk to Richard to dissect that. We all work hard at RCR. I don't think you can point the finger at one guy, at one point of that team, and say this is the reason. I think you have to look at the whole spectrum. It's a whole team effort.

"Richard Childress has always been a force at restrictor-plate race tracks, but I want to be that way everywhere we go. It didn't jump-start us in February. Last February was great, and the honeymoon isn't over yet, but it didn't translate to the rest of the season."

Although 1998 didn't turn out the way Earnhardt had hoped, he still had one glorious achievement to celebrate, which could stand as one of the crowning achievements in his NASCAR Winston Cup career—that is, the win at the Daytona 500.

"I woke up the next morning and still couldn't believe I had won the Daytona 500," Earnhardt recalled. "The disappointment you go through and the chapters of your life each year of the race, to finally win this race, well, this is big.

"I wish every race driver that ever runs Daytona could feel what I felt in Victory Lane. That's a feeling I know they have worked so hard for. It's a shame they can't have a lot of winners, but that is what makes the Daytona 500 so elusive.

"That's one of the greatest feelings in your life, to work that many years and come so close and be so dominant and finally win that race. It's an accomplishment I won't forget."

It's also a victory that stands the test of time.

"When Darrell Waltrip and I are sitting around old on the front porch and he talks about his three championships and his Daytona 500 victory, I can too," Earnhardt said. "We used to pick on each other. I'd say, 'Well, Darrell, I've won seven championships.' He'd say, 'Yeah, but you ain't ever won the Daytona 500.'

"He can't say that anymore."

No longer could Earnhardt be asked the question, "Are you ever going to win the Daytona 500?"

21

DOMINATOR OF THE DECADE—

Dale Earnhardt vs. Jeff Gordon: Who Is the Driver of the 1990s?

BY JON FITZSIMMONS AND STEVE ZEPEZAUER

Circle Track December 1999

When it comes to people who dominated NASCAR Winston Cup racing in the 1990s, Dale Earnhardt and Jeff Gordon are the championship dominators. Earnhardt has claimed four championships while Gordon claimed three (so far). The only other people to squeak in a championship during the 1990s was Terry Labonte in 1996 and the late Alan Kulwicki in 1992. Numerous showdowns on the track speak volumes about Gordon and Earnhardt, perhaps the most memorable being the 1999 Daytona 500, where Earnhardt left a subtle reminder in the form of tire mark donuts on Gordon's car following the checkers.

While the topic remains infinitely debatable, the fact remains that the line is blurry in regards to who was the dominant force of the 1990s. Some people say Earnhardt; others say Gordon. We say…

Statistically, one could argue that Earnhardt earned the reputation as Dominator of the Decade—four championships in one decade compared to Gordon's three (as of deadline for this issue, Dale Jarrett is the heavy favorite to win the 1999 title). Statistically though, Gordon has won more races, has a higher average point standing (albeit it includes three fewer years of racing in the 1990s than Earnhardt), and has won more money. The winnings aspect can be discounted to some degree, considering the purses are bigger in 1999 than in 1990.

Gordon's 2.8 average in the final point standings and the 46 wins (up to the 1999 Brickyard 400) can't be denied. Neither can the fact that Gordon has won two Brickyard 400s and two Daytona 500s in the decade. Earnhardt supporters can point to the Intimidator's total dominance in the first half of the 1990s, including four championships in five years. Regardless of what side of the Earnhardt/Gordon fence you're on, there's enough fodder to keep the debate going for decades to come.

Dale Earnhardt's Results 1990–1999

Year	Races	Wins	Total Money Won	Point Standing
1990	29	9	$3,083,056	1st
1991	29	4	$2,396,685	1st
1992	29	1	$915,463	12th
1993	30	6	$3,353,789	1st
1994	31	4	$3,300,733	1st
1995	31	5	$3,154,241	2nd
1996	31	2	$2,285,926	4th
1997	32	0	$2,151,909	5th
1998	33	1	$2,990,749	8th
1999	20/34*	1*	$1,883,749*	7th*
totals	315	33	$25,516,300	4.2 (average)

Jeff Gordon's Results 1992–1999

Year	Races	Wins	Total Money Won	Point Standing
1993	30	0	$765,168	4th
1994	31	2	$1,779,523	8th
1995	31	7	$4,347,343	1st
1996	31	10	$3,428,485	2nd
1997	32	10	$6,375,658	1st
1998	33	13	$9,306,584	1st
1999	20/34*	4*	$4,179,241*	6th*
totals	208	46	$30,182,002	3.3 (average)

* Not a final result. Race 20 was the Brickyard 400 on August 7.

Car #3 and car #24 have won more than 75 Winston Cup races between them in the 1990s. *Harold Hinson*

Gordon or Earnhardt: Who is dominator of the 1990s? *Nigel Kinrade*

RUNNING THE INTIMIDATOR'S BUSINESS

Don Hawk Q&A

22

BY BOB MYERS
Circle Track March 2000

Circle Track: *As president of Dale Earnhardt Inc. (DEI), what are your responsibilities?*

Don Hawk: I manage the affairs—endorsements, sponsorship packages, contracts, and personal matters—of Dale Earnhardt and Dale Earnhardt Jr., as well as those of DEI drivers Steve Park and Ron Hornaday. I make sure that when Earnhardt straps into his famed No. 3 Chevrolet, he doesn't have to worry about business. My objective is that when Earnhardt retires as a driver, he and his wife, Teresa, won't ever have to worry about money.

CT: *What is DEI comprised of?*

Hawk: Four race teams, two for Dale Jr. in Busch and Winston Cup, Park's in Winston Cup, and Hornaday's in the NASCAR Craftsman Truck Series; Dale Earnhardt Chevrolet; an automobile leasing company; rental and lease property; Black Angus cattle; and a poultry business through which we raise chickens for Perdue. We are active in the souvenir, apparel, and garment business and serve as

Paul Melhado

Hawk manages the endorsements, sponsorship packages, contracts, and personal matters of two of racing's biggest personalities: Dale Earnhardt and his son, Dale Earnhardt Jr. *Warren Melhado*

spokesman for some of the world's finest products, such as Coca-Cola, Budweiser, Burger King, Snap-on Tools, and Chevrolet. Teresa is chairman, and Earnhardt is chief executive officer.

CT: *How large are the new DEI shops, which resemble the Taj Mahal, near Mooresville, North Carolina?*

Hawk: Altogether, more than 300,000 square feet. The main building is 202,000 square feet, which includes a new, 82,000-square-foot addition to house Dale Jr.'s Winston Cup team.

CT: *How much did the facility cost?*

Hawk: Way too much. [A *Circle Track* writer was served a Diet Coke in a glass of fine crystal—a touch of Teresa class.]

CT: *Do you or Earnhardt make most of the major decisions?*

Hawk: It's a threesome—Dale, Teresa, and Don. Teresa is very sharp and is involved in almost every major decision.

CT: *Do you have a vote?*

Hawk: Yes, and that's unique in my relationship with Dale and Teresa. They respect the fact that I sometimes disagree. Ben Franklin said that when two people in business always agree, one

Dale Earnhardt, his wife Teresa, and Hawk comprise the front office of DEI. "Teresa is very sharp and is involved in almost every major decision," says Hawk. *Paul Melhado*

of them is unnecessary. If I sit here as a yes man and let the Intimidator intimidate me, I'm not a very valuable employee.

CT: *How is Earnhardt to work for?*

Hawk: He's a bear—very tough and intimidating. My job is to turn lemons into lemonade so we can both drink and be merry.

CT: *Can you give me a ballpark figure on how big the business is?*

Hawk: It's a privately held company, so I can't release any numbers. The number *Forbes* magazine used is $100 million.

CT: *Does Earnhardt make more money off the track than on?*

Hawk: That's a fair assessment. He makes a lot driving a race car and more than that off the track. [*Forbes* reports that Earnhardt earned $24 million in 1998, third behind Michael Jordan's $69 million and Tiger Woods' $27 million among pro athletes. According to the magazine, Jeff Gordon grossed $13 million.]

CT: *Do DEI business matters detract from Earnhardt's commitments as driver for Richard Childress Racing?*

Hawk: Not a bit. People talk about that a lot, and there was a time when Childress wasn't sure about that. When Dale buckles up, he's ready to race. He can focus in and focus out—like flipping a switch.

CT: *Why has the business side of racing gone into orbit in the past few years?*

Hawk: I think that NASCAR Winston Cup racing is on a parallel with the NBA and NFL. Our growth curve has been in the late 1980s and 1990s. We're seeing growth in merchandising, marketing, and strategic alliances with large corporations. There are corporations, led by Fortune 50 and Fortune 500 companies, involved in our sport that 10 years ago wouldn't touch it. A lot of credit goes to NASCAR marketing executives Brian France and George Pyne.

CT: *In terms of business, are you aggressive?*

Hawk: Extremely. I am a type A–driven personality. Oftentimes, my aggressiveness is seen as hostility and as a negative. There's always some stress involved in negotiating a deal, large or small. One thing I'll never do is compromise my integrity and honesty. And when I'm finished with a negotiation, I'm done. There are no hard feelings or bitterness on my part.

Hawk's main objective is to make sure that when Earnhardt is behind the wheel of the famous #3 Chevrolet, he is focused on the business of driving. *Harold Hinson*

CT: *What's the biggest business deal you've made at DEI?*
Hawk: We sold one of our companies, Sports Image, which we had bought for $6 million, to Action Performance for $24 million in cash and $6 million in stock. That stock is worth $13 million now [as of September 1999]. That's a handsome profit. And Action Performance has to pay Dale a guaranteed sum of money through 2013. Additionally, the Earnhardts' endorsement and sponsorship contracts with Budweiser are pretty big deals.

CT: *As Earnhardt's number of wins has declined, have his souvenir sales dropped?*
Hawk: No, his sales are as healthy as they ever were. There is a misconception about Dale's '99 season. He's won at Talladega and Bristol [through the first 26 races], but people forget that he also won a 125-miler at Daytona [and was second in the 500 and 400] and the IROC championship, winning three of the four races. I think all this is a sign that Dale Earnhardt is not done.

CT: *How do Dale Jr.'s souvenir sales compare with his daddy's?*
Hawk: Sometimes Dale Jr. is very close. At Richmond in September 1999, for example, Dale Jr. won the Busch race and finished 10th in the Winston Cup race. Dale finished sixth in the big race, and his souvenir sales edged his son's by a nose.

CT: *How do Big Earnhardt's sales compare with Jeff Gordon's?*
Hawk: I'd say now Gordon is a couple of percentage points ahead of Dale at some tracks, and at others, we outsell him. What's happening now is that Tony Stewart is in the mix. It's a guess, but I think Stewart has taken a piece of Gordon's sales, because both are young and run up front. We lose our sales of Dale Earnhardt merchandise to Dale Jr., but the proceeds don't leave the company.

CT: *How much longer do you think Dale will drive?*

Hawk: He has been presented a new contract to drive for Childress through 2002, with a one-year option, but he hasn't signed it [as of late September]. Dale Jr. has asked his dad to drive for DEI as his teammate, and he's considering it. That's not to say Dale won't drive for Childress or that Steve Park would lose his ride. We'd add a third Winston Cup car.

CT: *What was your first job in racing?*

Hawk: My first full-time job was with Alan Kulwicki in 1992, the year he won the Winston Cup championship. If Alan [killed in a 1993 plane crash] had lived, I probably wouldn't be working for Dale Earnhardt. Actually, my first offer came from Joe Gibbs when he was putting together his Winston Cup team [in 1991]. But it didn't materialize because his sponsorship had not been finalized at the time.

CT: *How long have you worked for DEI, and how did you get connected with Earnhardt?*

Hawk: Seven years. At Dover, three months after Alan's death, Dale stuck his hand out the window of his race car parked in the garage and called me over. I was working for Geoffrey Bodine, who had bought Alan's team. I asked how he was running. He said he didn't call me over there to ask how he was running; he asked if I would go to work for him. I told him we could talk about it, but that I wasn't in any hurry to change jobs. His second question was how much money was I making. Then he said to meet him after practice. Meanwhile, Richard Childress told me he had recommended me to Dale, that he needed somebody to run his business. After practice, Dale walked up behind me, grabbed me by the shirt collar, took me to his hauler, and gave me three phone numbers. He said he was going on a hunting trip and that I should call him within three days. Obviously, I called.

CT: *What were your interests in high school and college?*

Hawk: Until 10th grade, I didn't care about school or whether I learned anything. I was a renegade, a loose cannon who spent a lot time in the principal's office. I had absolutely no focus. Then I had a [born-again] religious experience. I went with a bunch of kids to a revival camp meeting. I made the decision to change my life and that Christ was going to rule my life because I was an arrogant son of a gun.

CT: *Were you interested in NASCAR and Winston Cup at an early age?*

Hawk: Not really, just dirt track racing.

CT: *How long have you been married to Cyndee?*

Hawk: Going on 20 years. It's one of the finest things that has happened to me. It's a good balance. I'm wound to the max; she's very low-key, laid-back, and methodical. And she's smarter than I am.

CT: *How many children do you have, and what are they doing?*

Hawk: Four—Jessica, 17; Joy, 15; Jennifer, 14; and John, 13. They're all in Christian school in Concord [North Carolina]. Jessica is a senior and will go to college next year. She's looking at Grove City in Pennsylvania.

Self-proclaimed as a "type A-driven personality," Hawk's aggressive business tactics have contributed greatly to the fame and fortune of DEI. *Harold Hinson*

CT: *Where do you reside?*

Hawk: About 5 miles from DEI on 7.5 acres. We have a fishing pond and a swimming pool. My philosophy is, as my teenagers grow up, I want all the kids to hang at my house, because I want to be able to control the environment.

CT: *What's your favorite leisure activity?*

Hawk: Golf. I'm playing more often now and getting my game together.

CT: *Age 44 and a native of Pottstown, Pennsylvania, you have a bachelor of science degree in Bible with a minor in history from Philadelphia College [now University] of Bible. Are you an ordained minister or teacher?*

Hawk: I am not an ordained minister, but I am certified to teach Bible and Christian Education. Also, I'm on the board of trustees of my alma mater, and that's quite an honor.

CT: *What's your faith?*

Hawk: Lutheran. I grew up as an independent. My wife, Cyndee, is Lutheran, and after we were married, I didn't think it was fair to ask her to become an independent, especially because I travel 28 weekends or more a year.

CT: *Do members of the Winston Cup racing family come to you with questions and problems regarding faith?*

Hawk: Yes, I've had drivers, owners, sponsors, licensees, and licensors talk to me about life, death, babies, ethics, and morals. I think that's part of why I am where I am.

CT: *Given this type of training, why and how did you get into racing?*

Hawk: I've been around racing most of my life. When I was little, my parents and two older sisters took me to Dorney Park Speedway, a 1/8-mile dirt. I also hung around Nazareth [Pennsylvania] Speedway, which was owned by my brother-in-law before Roger Penske bought it. I developed a love for cars and speed. When I was 17, I got into the automobile business, prepping new cars at a dealership in Allentown, Pennsylvania. After college, most people expected me to teach theology, but I returned to the automobile business and worked my way up to director of parts and services. I didn't feel comfortable at that time with being a preacher or a teacher. I spent 18 years in the car business.

CT: *In demand as a speaker, what sorts of groups do you speak to?*
Hawk: I try to do a lot of charities and big business, college and university business schools, churches, and religious organizations. Not all of my education came in a classroom—it came in the street, and street-smart is a little better.

CT: *Aren't you a workaholic?*
Hawk: Yeah, and to my shame, sometimes I work too much. If I had it to do over, I'd work less and spend more time with my family. But I get paid a lot of money for what I do, and I do it well. I think Dale and Teresa would allow me to have more time with my family; I just haven't taken it. I'm just starting to free myself up a little bit.

CT: *In addition to receiving numerous honors and awards, including Ford Motor Company's Distinguished Service Award seven times when you were with Ford dealerships, major publications have ranked you among the top-25 most influential people in NASCAR. In what ways are you influential?*
Hawk: It's an honor to be on that list. I think part of that is when I came to DEI, Dale's picture wasn't on cereal boxes, he wasn't with beverage, fast-food, or die-cast car companies. We brought a new business approach to DEI, but beyond that, I knew that my strategic alliance partner was NASCAR. I told Dale I could see huge growth in the sport over the long haul.

CT: *Did you negotiate Dale Jr.'s multimillion-dollar Winston Cup contract with Budweiser?*
Hawk: Absolutely.

CT: *Is the six-year contract valued at $10 million per year as speculation indicates?*
Hawk: It's for every bit of that amount. I can tell you that if we don't have $10 million a year on the No. 8 car, I'm not doing my job.

CT: *What's the going rate of a first-class, primary Winston Cup sponsorship these days?*
Hawk: To be competitive, between $6 and $8 million per year.

CT: *Do primary sponsors demand more for their dollar these days?*
Hawk: By far. A pet peeve of mine is that primary sponsors used to come into this sport to advertise. Now they come not only to advertise and demand greater performance, but they also want to make money off souvenirs, drivers' appearances, and that kind of thing. I think that's wrong. As the price goes up, sponsors deserve the right to make more demands, but only in proportion.

CT: *Isn't Dale Jr. one of the hottest properties?*
Hawk: Yes, already he's one of the top three.

CT: *Has Dale Jr.'s success on and off the track surprised you?*
Hawk: It came quicker than I thought it would. What has surprised me the most is his ability to handle his success, especially the media.

Hawk was instrumental in negotiating the sponsorship deal with Budweiser for Dale Jr.'s #8 Chevrolet. *Paul Melhado*

CT: *Is marketability almost as important as driving ability for today's drivers?*

Hawk: Absolutely. If you don't have it, you're not going to sell. It's that simple.

CT: *From a business standpoint, what would be your advice to up-and-coming drivers with obvious talent?*

Hawk: Find someone who is going to take care of you, treat you with respect, integrity, and honesty, and say no to the first guy who offers you a lot of money. *Life's Little Instruction Book* says to say no to something every day.

CT: *Anything you wish to add?*

Hawk: To me, I took the road less traveled. I decided to come to a sport that I knew nothing about and take a crash course in it. If tomorrow Dale and Teresa told me it was over, I could live with that and be happy. If they wanted to extend my contract five years, I'd say, "Go get the cake and light the candles." I can't emphasize enough that I'm 100 percent happy. Dale and I have a lot more money in the bank, a lot more friends in the world, and deeper relationships with corporate America than we had when we met that day at Dover.

23

DON'T COUNT ME OUT

Dale Earnhardt Is Still a Force in Winston Cup

BY BRUCE MARTIN
Stock Car Racing June 2000

Dale Earnhardt was named NASCAR Driver of the Decade for the 1990s and some will argue he may be the greatest Winston Cup driver of all time. In 2000, he is ready to continue that legacy into the new century.

The driver who began the 1990s with four Winston Cup titles in the first five seasons was ready to lay claim as the driver of the century. But Earnhardt began to falter beginning in 1996 when he was involved in the worst crash of his career at Talladega in the DieHard 500.

What followed was a losing streak that reached 59 races before it finally snapped when he won the 1998 Daytona 500, ending 20 years trying to win NASCAR Winston Cup racing's crown jewel. But another winless streak followed before it was snapped in the DieHard 500 at Talladega in 1999—the second win in a 101-race span for the once dominating driver.

Earnhardt wasn't through. He bumped his way to victory at Bristol Motor Speedway in August when he knocked Terry Labonte out of the way on the final lap in vintage Earnhardt style. It was his first multi-win season since 1996. He sealed a fantastic season with the third win of the year in the Winston 500 at Talladega on October 17.

It signaled to the doubters that Dale Earnhardt was back.

"Hell, in my mind, I had never left," Earnhardt says, displaying his cocky grin. "I never went anywhere. I have always been here. I'll be here until I'm gone. After this deal with Richard Childress is over, I'll probably look at retirement. I'm not going to make any bones about it or set a time deal on it."

As Earnhardt entered the 2000 season, however, he was recovering from neck surgery to correct some pain and discomfort. It gave the bigger-than-life racing hero a chance to sit back and reflect, while also preparing to look ahead for the final stretch of his fabulous career.

Harold Hinson

"I enjoyed the time off," Earnhardt says. "I healed up and got stronger again. I'm in good shape. I'm ready to race. Our goal to win that eighth championship with Richard Childress and GM Goodwrench Service is what is motivating me now. I just signed a brand-new contract with Richard Childress and GM Goodwrench. It's good through the year 2003 with options for more, so there's no retirement clause in there or anything, so we're in good shape."

Winning the Brickyard 400 in 1995 and then the Daytona 500 in 1998 was a huge sense of relief, but not the final exclamation point of his career. Earnhardt believes that will only come with an eighth Winston Cup title.

The 2000 season has given Earnhardt renewed hope and the confidence that he can continue as one of the sport's dominant faces. Although he struggled for a portion of the 1990s, he believes there were a variety of factors that led to that situation.

"I think what happened is the competition kept growing," Earnhardt says. "We didn't probably do the right things—Richard Childress and us. Finally, we have turned it around and put things back in perspective and I think our race teams are back on track and we have the focus we need on a lot of things.

"I think the feel I have for a race car is different than probably somebody that is younger because of my years in racing. I think Kevin Hamlin and I have turned the corner on having a great relationship and

Jeff Gordon's success in the 1990s was a big obstacle in Earnhardt's quest for his record-breaking eighth Winston Cup title. *Warren Melhado*

making it work better. Our communication is coming together a lot better than what I had before. I think that is a big part of it, making our team turn the corner. There are a lot of things like that."

When Earnhardt won the Daytona 500 in 1998, it was thought to be the start of his breakthrough year after several seasons of struggle. But the breakout came in 1999 with his victory at Talladega, solidifying his reputation as the master of restrictor-plate racing.

"I try to work the car in practice and qualifying and get it to where it really runs good in the draft, in the air," Earnhardt says. "If I can, I'm more confident in it and I run better in the race. I learn a lot in drafting. Restrictor plates changed the drafting when it came on board and as it went through the years; it changes even more with bodies and air and spoilers and everything that you may or may not do. That's important. I've been good at using the air."

The Bristol victory last August let the racing world know in no uncertain terms he still had that vintage Earnhardt style of using fear and intimidation, and a good hard knock, to his advantage.

"The whole race was overshadowed by the end of the race because we ran so good that night," Earnhardt says. "We ran from the back to the front and even on the back straightaway; we were running good."

Part of the reason why Earnhardt and Childress have lasted together for two decades is their ability to work together. Earnhardt has always been more than just a driver at Richard Childress Racing. He has played a vital role in the development of the racing operation and his input is invaluable.

"I have just as much confidence in him winning that eighth championship now as I ever have," Childress says. "If we get our stuff right, we can do that.

"This race team—we are building it back to where it was. Dale was hurt a couple of years. With the whole package of good cars and our team, you will see the vintage Dale Earnhardt for several more years. As long as Dale Earnhardt is healthy and he is competitive, he wants to keep racing."

Earnhardt also has additional motivation in 2000. His son, two-time Busch Series champion Dale Earnhardt Jr., is driving the Budweiser-sponsored car for Dale Earnhardt, Inc. Steve Park

Earnhardt signed a new contract to race for Richard Childress into the 21st century. *Nigel Kinrade*

Harold Hinson

is the other DEI driver in the Pennzoil Chevrolet Monte Carlo.

"Before, I was Dale Earnhardt," he says. "Now, I'm Dale Earnhardt Jr.'s dad. I think that put some light on my age and the time I've had in this sport. Everyone realizes I've gotten older. Fans and people have talked retirement or said I was through with racing. They've seen the greats come and go in racing and not win in the latter years—A.J. Foyt, Richard Petty. But still, it was hard for those guys to quit racing. It's like Darrell Waltrip right now. He's talking retirement and it has to be tough for him as a driver to stop wanting to drive."

One of the most special moments in Earnhardt's life came from a father-and-son encounter. It was in the early 1970s when he raced his father, the late Ralph Earnhardt, at Metrolina Speedway in Charlotte.

"I did race against my dad at Charlotte in the dirt one time, but I never really did it in a Sportsman car like he had," Earnhardt says. "The one thing I probably would liked to have done that I didn't get to do was drive for my dad. The year he died, he was talking with Mom about putting me in the car the next year rather than having other people drive for him. I didn't find it out until after he passed away and my mom and Uncle Bud was telling me about it. That would have been great. I would have gotten my ass chewed out a lot, but that would have been great. I think that would have been a lot of fun."

Earnhardt has grown to appreciate and enjoy racing against his son in Winston Cup competition this season. "I'm proud of it, but I still want him to learn and do the right things. He gets excited racing around me and I see it. He wants to beat me when he just needs to really race smart. He'll end up beating me like he will anybody else. I'm just proud to see him race.

"Dale Jr. and Kelley and Kerry, all of us have a pretty good relationship. I want the kids to be themselves. I want them to be able to excel in what they can do. Dale Jr. really wants to race. I think he was paying better attention than what I thought."

One of those special moments came in last year's IROC race at Michigan. Both cars came out of the fourth turn side by side with the checkered flag in sight. "That was a fun day, it really was," Earnhardt recalls. "He was patient and I could see he wanted to be the leader, wanted to be up front, but still, he was patient. Finally, he made that last move to the end there and I was lucky to still beat him. That was fun. That was classic. I think he used his head a lot. If he keeps doing that, I think he'll make it as a racer."

With success comes media attention. Earnhardt patiently answers questions following the 1998 Daytona 500. *Nigel Kinrade*

While the rest of the field follows the pace car, Earnhardt gets service from the GM Goodwrench crew. *Harold Hinson*

Earnhardt started his defense of the '99 IROC title by winning the first event of the 2000 season at Daytona. *Harold Hinson*

Father Dale feels bad that his racing career often kept him from spending quality time with his kids while they were growing up. By sharing a relationship in racing, it has helped the two bond and connect with each other.

"As we grew up and I was racing and they had school, they missed a lot of time with me," Earnhardt says. "Now, when they are up and going and have all this stuff going, I call and want to do stuff with them and I miss a lot of time with them. We have to schedule time together. That's pretty neat.

"Yes, it did bother me to leave them at home, but it was smart to leave them there and get their education. I have some hard-to-beat pictures of them as kids back in 1986 and 1987 and 12 years later, Dale Jr. is winning championships. It's hard to believe seeing those pictures."

During that time, Dale Earnhardt Jr. used to daydream of what it would be like to beat his dad on the race track. This year, he will get his big chance.

"It's just like kids growing up wanting to be Michael Jordan or wanting to be Dale Earnhardt," Dale Jr. says. "When I was growing up, I wanted to be Dale Earnhardt. He was the champion. He was my dad, but he was the guy who was winning and he was the guy I wanted to be. I always thought about driving his car and stuff, so it was kind of cool to be able to do that."

Earnhardt is a proud man in many respects. One of the things that brings him a tremendous amount of pride is the racing empire that he has built, along with his wife, Teresa.

"I'm proud of what we have accomplished, Teresa and myself and all the folks that supported us and worked with us," Earnhardt says. "I'm proud of my family and the way they have grown up. They may not be the brightest kids, but they are good kids. They have done well.

"Teresa helped me. I have the common street sense and she has the book smarts. We are fortunate. She has a great family and mom and dad. Her dad is the type of guy who can take chicken

The thrill of victory. Champagne goes spraying into the Tennessee night air in Bristol's Victory Lane. *Sam Sharpe*

shit and turn it into chicken salad. He is good at making ends meet and making things happen. He takes things and turns them around. My dad was like that. He was an innovator. He'd take a little of nothing and make a good race car out of it and win races. We both came from a hard-luck background where we take what we have and make the best of it.

"People always talk about us not having a lot. Shoot, I thought we were rich. I never thought we didn't have a lot. I thought we were always rich. I never wanted for anything. Teresa and myself grew up that way and I think we've been fortunate, very fortunate."

For Dale Earnhardt, it has truly been a wonderful life.

Family is important to Earnhardt, and having son Dale Jr. *(left)* in the Winston Cup will represent a new challenge. Wife Teresa *(right)* has been right there for her husband every step of the way. *Staff photo*

He's still the man to many—this young fan wears a proud allegiance. *Nigel Kinrade*

24

THE MAN WHO WOULD BE KING

Dale Earnhardt's Reason for Racing

BY RON LEMASTERS JR.
Stock Car Racing November 2000

Six years ago, it seemed as if Dale Earnhardt was a hair's breadth away from rewriting NASCAR history, dethroning the King of Stock Car Racing, and crowning the epoch of Earnhardt with his eighth Winston Cup title.

There were T-shirts made touting the "7&7" status of both Earnhardt and the King. Talk was rampant about how soon there would be a new all-time leader in championships. Earnhardt was, as they say, a mortal lock for 8, even 10 titles by the time he was through.

Alas, history is not so easily made, nor are kings willing to give up their crowns. The fact that Dale Earnhardt has not won that pivotal eighth title is a matter of record, and each year that passes speeds the Man Who Would Be King that much further away from his appointed date with a legacy surpassed by no other driver.

How did it happen?

That's easy enough to answer. In the six years since his last title in 1994, a new Crown Prince has come along in the person of Jeff Gordon, who has won half those six titles. Terry Labonte won in 1996 and Dale Jarrett won in 1999, each taking a turn at the top of the heap and robbing Earnhardt of another chance. At 49, it's not like he has that many years left at the peak of his game.

There are also the injuries he's suffered. Earnhardt went through three or four seasons with nagging injuries and other maladies that had nothing to do with race cars on race tracks—broken bones, a mysterious fainting spell just before the Southern 500 one year, and surgery to correct some nerve damage.

It's not as if Earnhardt hasn't thought about passing Petty. He just didn't talk about it much before he tied the King with his seventh championship in 1994.

"In stock car racing, Richard Petty will always be the King," Earnhardt was fond of saying. "No one will ever outdo him."

Sam Sharpe

In 1998, when he won the Daytona 500 for the first time, someone asked him if his career was complete. "Hell no," he said. "We want to win that eighth championship. That's what my life and career have been all about: winning championships. Nobody's ever won eight before, and that's what we're shooting for. I think we've got a great shot at it this year, and we'll keep going from there."

Of course, that was the year Jeff Gordon laid 13 victories and 26 top-five finishes on the field, and his 364-point victory margin over Mark Martin was the largest since Earnhardt's record 444-point bulge in 1994. Richard Petty won his seven titles in an era when stock car racing was a somewhat regional sport. Earnhardt's first, in 1980, came as the tide toward national appeal and acceptance was just beginning to run. The six titles since then—in 1986, 1987, 1990, 1991,

Has time, the enemy of all men, finally caught up with him? No one but Earnhardt can be sure. *Sam Sharpe*

1993, and 1994—have come in an age when the tiniest margin loomed large enough to drive a transporter through. The competition was better top to bottom, and several of Petty's toughest competitors raced sporadically, like David Pearson and Cale Yarborough.

"People didn't run for championships then," Petty said a few years ago in a newspaper interview. "They ran to win races. When the season was over, they added it up. There could have been three or four championships I could have won if I had been looking at it that way."

Dale Earnhardt Jr.: "I'm really excited about it, the opportunity for him to get his eighth championship. I'll be right there beside him if he does." *Paul Melhado*

Earnhardt has not had that luxury. These days, everything is about championships. There is no cherry-picking, and missing a race means that your team is in trouble or you're about out of money—the new official fuel of NASCAR.

Legendary *Charlotte Observer* reporter Tom Higgins once asked Petty if he could compare himself and Earnhardt. Petty said it wasn't that simple. "It's one of those deals where you can't compare," Petty said. "It's like Hank Aaron beating Babe Ruth. They did their things in different times. Same with me and Dale. Right now, he's the top dog going through here."

"I went through with Pearson, Cale Yarborough, Fred Lorenzen, and Bobby Allison. So where would that put Earnhardt if he had run against them? He'd be up there with them. But if you go back and try to figure out of all the drivers who's the best, there ain't no such thing."

What are Earnhardt's chances of catching Richard Petty and winning that eighth title? That question is moot if he should not manage to catch Bobby Labonte and Dale Jarrett in the race for the title this season. This 2000 season has been Earnhardt's best since at least 1996, when he finished fourth in the points behind Terry Labonte, Jeff Gordon, and Dale Jarrett. Since then, the best Earnhardt has finished in the points was fifth in 1997. In 1998 he slipped to eighth and was seventh in 1999.

This is his best chance to get the magical, mystical eighth title in the past three seasons. As of mid-August, he was third in points behind Bobby Labonte and defending champion Dale Jarrett, and he's running as well as anyone these days. In his heyday, Earnhardt won titles with bruising efficiency, stringing together top-five and top-10 finishes like popcorn on a holiday garland. By the time the season hit the home stretch, he was usually so firmly in control the race was for second place, the title all but won.

Gordon's three-titles-in-four-years domination of Winston Cup racing from 1995–98—doing Earnhardt's run of three titles in five years one better—set a new standard, one the old master has had a little trouble living up to. Certainly, he can still rumble with the best of them, but he's not the Intimidator of old anymore. What advantage his apparent willingness to do

"I think I've been a good driver the whole time," Earnhardt said of his recent woes. "I enjoy the pressure. I thrive on it." *Nigel Kinrade*

whatever was necessary—sometimes fair and sometimes smudging the line between right and wrong—gave him in those years, it has been erased by a combination of time and changing attitudes.

Has he lost the edge? Has time, the enemy of all men, finally caught up with him? No one but Earnhardt can be sure, but it's just possible that some of Earnhardt's best racing may be yet to come.

"I think I've been a good driver the whole time," Earnhardt said of his recent woes. "I enjoy the pressure. I thrive on it. I'm not one to go out and overdrive the corner because someone is pressuring me into the corner. I try not to do that."

His son, Dale Jr., sees that familiar killer instinct in his father this year.

"I'm really excited about it, the opportunity for him to get his eighth championship," Dale Jr. said recently. "I'll be right there beside him if he does. He just really got pumped up this year, not so much for the eighth title but just to be running good again. He is such a hard guy to figure out." What's not hard to figure is his largely unspoken desire to win number eight.

Is time running out on Earnhardt? Yes and no. He's signed to drive for Childress through the 2003 season, giving him at least another four shots at the record. In addition, he owns a two-car Winston Cup team. Both his son Dale Jr. and teammate Steve Park have won races this season, proving that those teams could, if necessary, be the ones Earnhardt uses to take the crown from Petty's head. "I've got four years of racing left, at least," Earnhardt said. "Who knows—I might even drive another car with my own team. I'm not ruling anything out."

The Chiefs

No less an authority than three-time NASCAR Winston Cup champion Darrell Waltrip, who has never been shy about sharing an opinion, spoke of championships as things accomplished by many, not by one. "Drivers win races; teams win championships," was the way he put it, and Earnhardt has made a great living off just that sort of thinking.

Earnhardt has utilized the talents of a number of crew chiefs on his way to seven titles. In his first title, Doug Richert was the wrench for the final 18 races after "Suitcase" Jake Elder quit in May. His next four came with Kirk Shelmerdine at the helm, and his final two were with Andy Petree. Current crew chief Kevin Hamlin took over from Larry McReynolds 14 races into the 1998 season.

"I've got a job and an opportunity to win that eighth championship," Earnhardt said, talking about his refocused RCR team. "That's what we're focusing on. That's what we're driving for. That's what we're working for. We're doing it with a more solid team than I've had in past years."

THE NUMBERS

Earnhardt has one of the most enviable records in NASCAR history to go along with the seven titles. Seventy-five victories, 275 top-five, and 420 top-10 finishes in 663 career starts (through Watkins Glen this year). In prize money alone, the King of Kannapolis has won more than $37 million in 26 seasons.

A closer look at the numbers reveals that Earnhardt is at his best in the midst of a point race. For instance, his last title came in 1994, and during that season he earned 25 top 10 finishes in 31 starts to go along with 4 victories and 20 Top-fives. That sort of consistency has become the hallmark of his career, and it is the model upon which all teams base their quest for the title. If you're around the front of the field, regardless of whether you win or not, the points will take care of themselves.

After 20 races, or a little more than halfway into 2000, Earnhardt was third in the points, 217 behind Bobby Labonte. He had 16 top 10s in those 20 races, 7 in the top five, and a victory over Labonte—by less than a foot—at Atlanta in March.

The race for the 2000 title—and possibly Earnhardt's eighth—boils down to the top three drivers, according to Earnhardt.

"I think the guys you've got to be concerned with are Jarrett and Labonte," he said. "Sure, all of them can be a threat. All of them that are in the top 10 can be a threat for the championship. It took half a season to get where you're at now. It'll take another half of a season to get somewhere else. I do think the No. 88 and No. 18 are the guys we're going to be racing for that championship."

Nobody has been in more point battles than Earnhardt, with the possible exception of Mark Martin, who has yet to win any of those battles. When the season comes down to crunch time, and every point is precious, Earnhardt has a particular way of dealing with it.

"If you're counting points in October in the races, you might be counting down. At this point, I'm not counting them off. I'm just waiting until it's closer to the end, and then I'll add them up."

Ask Earnhardt which one of the seven titles he's won is his favorite, and the answer is vintage Intimidator. "All of them have been hard," he said. "Winning your first championship is hard. Winning the one after that is hard. Ask anybody—Dale Jarrett, Bill Elliott, or Rusty Wallace. You don't go out and just do it again."

Winning an eighth—despite the historical significance—is no easier or harder than the first one, according to Earnhardt. "Winning the eighth championship is the next one. It's our next goal. That's our next focus. Yeah, it's important. We're competitive, and we have the team to race for it.

"I don't weigh if I want to win this race, or put this race against that race or this championship against that championship. I've really enjoyed them all. Again, I'm so damn fortunate to grow up in a family that raced. To get into it and accomplish what I have in my life, it's pretty awesome. I'm excited about what I do. I'm not content with not winning. If somebody tells you I'm riding my years out, they're not paying attention."

Kirk Shelmerdine. *Harold Hinson*

Andy Petree. *Floyd Coffman*

Coming off a tight battle with Gordon in the 1995 points chase, Earnhardt hit a major obstacle: injury. During a battle through the tri-oval at Talladega Superspeedway, Sterling Marlin nudged Earnhardt's car in the right rear and sent him catapulting into the outside wall. The heavy impact cracked his sternum, the bone that holds the ribs together. Ken Schrader piled in at high speed, having nowhere else to go, and that shot broke Earnhardt's collarbone.

Those injuries, along with back pain similar to that which shelved Mark Martin the entire off-season, combined to lay him low in both 1997 and 1998.

"I may have been hurt for the last two years and working with pain and stuff and didn't realize it until it got worse and worse," Earnhardt said. "I had to have something to do about it. That's all in the past. It's over. We're healthy. The team is healthy. Everything is there. Now it's ours to capitalize on it, move forward, and get those top fives."

In 1998 Larry McReynolds came over to Earnhardt's camp from Robert Yates to replace Petree, who had left to form his own team. The pairing was thought to be the missing link Earnhardt needed to earn that elusive eighth title. But it was a bust, despite the fact that McReynolds helped Earnhardt win the Daytona 500 in 1998. By the middle of that season, McReynolds was swapped to the RCR team car of Mike Skinner, and Skinner's crew chief, Kevin Hamlin, was Earnhardt's new signal-caller. Hamlin, an unassuming Michigander, provided the magic that reminded Earnhardt of Shelmerdine.

"Kevin Hamlin is more my type and style of crew chief," Earnhardt said. "He's constantly thinking. He never quits on making it better. He's not the kind of guy who blows it off and kicks and screams. He's the kind of guy who sits down in here, listens to what I say, talks about what we're doing, and tries to make improvements.

"He's so much like Kirk. When I say something, he's already thinking it, and that's the way Kirk was. We're thinking a lot to the same note. It's been that way. I'm so happy Richard Childress made that change. I haven't regretted it one bit. I love Larry McReynolds to death as a person. We won the Daytona 500 together, but day in and day out, I work well with Kevin."

A new Chevrolet Monte Carlo and time have improved the sometimes-surly seven-time champ's outlook on life this season. "I think our whole program has come together from last year to this year. The new Monte Carlo has helped us on downforce and the driveability of the car. Kevin and I have got more time under our belts together. Our crew has been fine-tuned and moved around, and everybody really wants to be there. You're getting paid to be there, but they want to be there. I want to be there," Earnhardt said.

"Over the years I've driven for Richard Childress, I've been offered bigger deals, but I'm with Richard because I want to race, I want to win, I want to be with somebody who wants to win as badly as I want to. I feel like that is the package we have this year. We want to overcome. Our pit stops were not that good the first of the season, but we worked on that, and our pit stops are right there with everybody else's.

Larry McReynolds. *Sam Sharpe*

Kevin Hamlin. *Harold Hinson*

Richard Childress. *Paul Melhado*

25

DRIVING FOR A RECORD

BY BOB MYERS
Circle Track December 2000

The Intimidator Has Thousands Believing a Record Is Going to Fall

Paul Melhado

For more than a decade, Dale Earnhardt was the favorite to win each Daytona 500 and didn't. He would win practically everything else during SpeedWeeks, but not the 500. Then, in 1998, coming off his first winless NASCAR Winston Cup season in 16 years and with very little prerace fanfare and hype, Earnhardt won the missing jewel.

I have the same gut feeling about the prospects of Earnhardt winning a record eighth championship this season—well, I did in late July. One of the great disadvantages—maybe risks is a better word—of writing for a magazine is lead time. That is, more than two months are required to get this from me to you. Situations can change as rapidly as the weather, and I can wind up looking like a donkey. So I can't know about October.

But at this writing, Earnhardt is stirring my juices and those of his army of thousands who want to see him break the seven-title tie with King Richard Petty. Earnhardt seemed to have the best opportunity to accomplish that feat since he was runner-up to Jeff Gordon in 1995, and the Intimidator, along with Richard Childress' Goodwrench Chevrolet team, once Winston Cup's benchmark, appear to be championship material again. The most important ingredient, consistency, has been restored. In the final seven of the

Paul Melhado

first 19 races, Earnhardt ranked no lower than third in the standings; in five straight races, he was second.

Given that Earnhardt is 49 years old and hasn't won a title in five years, he may not be the favorite. There was no hullabaloo surrounding him in late July as there was in the early 1990s when he was champion four of the first five years of the decade. The remarkable success of rookie Dale Earnhardt Jr. has shifted the focus on Earnhardt more as a proud father than as a driver, and there's another offspring, Kerry, 30, or "Middle E.," rising through the ranks. "I've had my share of the press," Earnhardt says. "It's time for somebody else."

Intimidating the Odds

Of course, the odds are against Earnhardt winning another championship. No one, certainly in the sport's modern era, has won it all at age 49. The late Lee Petty was 45 when he won his third championship in 1959. Bobby Allison was 45 when he won his first and only in 1983. Richard Petty won his last of seven at age 42 in 1979. David Pearson was 34 in 1969, Buck Baker 38 in 1957, and Cale Yarborough 38 in 1978. Among active drivers, Darrell Waltrip was 38 when he won his third championship in 1985, Terry Labonte 40 in 1996, and Dale Jarrett 43 in 1999. Earnhardt was 43 in 1994.

But Earnhardt is a master at beating the odds. Some wrote him off in 1992 when he won only one race and plummeted to 12th in the points chase, the first time he had finished out of the top 10 in nine years. Then he rebounded to consecutive titles and has finished no lower than eighth in the standings since. Considering his record, anything below second or third is a bad year.

And Earnhardt continues to win races, though not as many. Last year, at age 48, he logged three wins and three seconds. Few drivers, if any, have won a Winston Cup race past age 47. Lee Petty in 1961 and Richard Petty in 1984 were 47. Bobby Allison was 50 when he won the Daytona 500 in 1988, a few months before his career-ending accident. Buck Baker won the Southern 500 at 45 in 1964, when most people felt he was over the hill. Yarborough won for the final time at age 45.

Earnhardt and Childress share a never-ending desire to be the best. It is this shared philosophy that has kept the two partnered for so long. "Over the years that I've driven for Childress," says Earnhardt, "I've been offered bigger deals, but I've stayed because I'm with somebody who wants to win as badly as I do." *Nigel Kinrade*

Critics have claimed that Earnhardt has allowed too many distractions, including starting a race team with wife Teresa and mentoring son Dale Jr., to hamper his driving, but Earnhardt maintains that it has never been a problem. Besides, Earnhardt's outside projects have been hugely successful, with Dale Earnhardt Inc. claiming three Winston Cup points wins so far this season—two by Dale Jr. *David Ryan*

No one took charge of the points race in the first 19 races of 2000. No one was choking on Jeff Gordon's exhaust. Bobby Labonte led the points standings for 16 races, but by less than 100 points most of the time. There was no clear favorite. Dale Jarrett, definitely a strong contender to repeat, had more top fives and top 10s than Labonte, but the latter was more consistent. Sophomore sensation Tony Stewart was winless and 10th in points after the first dozen races, then won three times and soared to a

Age, injury, and the rise of family members into racing may have some people believing Earnhardt's days of winning are over, but that is simply not true. "I'm not content with not winning," The Intimidator quickly asserts. "If somebody tells you I'm riding out my years, they're not paying attention." *Nigel Kinrade*

sixth ranking. Other top 10 contenders—Jeff Burton, Rusty Wallace, Ward Burton, Jeff Gordon, Mark Martin, and Ricky Rudd—were up and down. Only 268 points separated first and fifth place. Earnhardt has as good a shot as anybody.

Why? Two prime reasons: Earnhardt has more confidence in crew chief Kevin Hamlin and his team, and he is healthy.

"I think our whole program has come together this year," Earnhardt says. "The team is more solid than it has been in past years. Kevin is more my type and style of crew chief. He's a man of few words and is constantly thinking. He's not one to blow up, kick, and scream. He sits down and listens to what I say and tries to make improvements. He never quits. He's so much like Kirk Shelmerdine [former crew chief who led Earnhardt to 41 victories and four titles at RCR through 1992]. I love Larry McReynolds as a person. We won the Daytona 500 together. But I work better with Kevin.

"The new Monte Carlo has helped us on downforce," Earnhardt continues. "Our crew has been fine-tuned and moved around to the point that everyone wants to be where he is. I want to be there. Over the years that I've driven for Childress, I've been offered bigger deals, but I've stayed

After struggling with pit stops in the first part of the 2000 season, crew chief Kevin Hamlin and the rest of the #3 team are firing off ultra-quick stops in an effort to get The Intimidator his eighth championship. *Nigel Kinrade*

because I'm with somebody who wants to win as badly as I do. I feel like that is the package we have this year. We want to overcome. Our pit stops were not good at the first of the season, but we've worked on that and gotten them to where they're as good as anybody's."

Earnhardt had surgery during the off-season on neck vertebrae that were damaged and subsequently aggravated in two crashes at Talladega, the latest in 1997, and says he is pain-free.

"I was hurting for the past two years and probably didn't realize it was affecting me until it got worse," Earnhardt says. "That's over with. I'm healthy. The team is healthy. The championship is there for us. It's ours to capitalize on—to move forward and get those top fives. If we can keep on doing what we've been doing and get close enough to put pressure on Bobby and Jarrett, we'll be in good shape.

"There's a lot of pressure on the leader as well as those trying to catch him. I enjoy pressure. I thrive on it, and that could work to my advantage in a showdown."

Earnhardt considers Labonte and Jarrett the top contenders. "Most of them in the top 10 [at 19 races] can be a threat," Earnhardt says, "but I think we're going to be racing the #18 [Labonte] and #88 [Jarrett] for the championship. I don't think we'll see one team excel and win several races. It seems to be the type of race where you've got to be up front in every race to be a contender. You can't afford many Bristols [finished 39th] and second Poconos [25th after nine consecutive top 10s]. I'm not counting points now [in August]. I'll wait until closer to the end, add them up, and see where we stand."

The Desire Still Burns

Earnhardt has started more than 600 Winston Cup races and with 75 wins has little left to conquer. Still, he has lost none of his fire and desire and gets excited when he cranks up.

"I'm just going through the motions in practice and preparations," he says. "I'm on an even plane when it comes to excitement, except on race day. The better shot I have at winning, the more excited I am. And I'm pretty excited this year. I'm not content with not winning. If somebody tells you I'm riding out my years, they're not paying attention."

It seems to be a fun year for Earnhardt. He seems more relaxed, and perhaps that's why some people are lulled into thinking he is not as serious about his career. He has derived a great deal of pleasure from the racing success of Dale Jr. and now the emergence of Kerry, but their careers are not a distraction.

"I'm proud and excited about my kids," Earnhardt says, "but that doesn't replace me winning races. That's what I'm driven to do. I get up in the mornings to race. I reckon I've got the best of two worlds—I'm a winning driver and a winning car owner. Look around. There aren't many people who have done that and done it the way I have.

"I'm so damned fortunate to have grown up in a family that raced. To get into it without a lot of means and resources and accomplish what I have, then have your kids come along with you is pretty awesome. I didn't force Dale Jr. to race—the kid wanted to. I'd just as soon that he'd have been a doctor or a preacher."

Of all Earnhardt has accomplished, he would appear greedy if he told us how much he wants the eighth championship. It would very likely be a last hurrah as a driver—a fitting crown to an illustrious career. The chances were very real in August.

If not this year, is there enough time? "I've got four years of racing left, at least," says Earnhardt. "If physically and mentally I can't do this, if I can't beat the next guy or win a race, then I'll stop. My reflexes and health will tell me when. I don't see that happening in the next three years."

On that point those who love and hate Dale Earnhardt might agree.

26

BY JIM MCLAURIN
Circle Track June 2001

EARNHARDT

A Nation Mourns the Loss of Its Greatest Stock Car Driver

When Dale Earnhardt crashed on the final lap of the Daytona 500 on February 18, 2001, causing his death, it set in motion a massive state of mourning that involved everyone in racing and many who knew him only as the driver of that imposing black No. 3 Chevy.

The outpouring of sympathy and support for Earnhardt's widow, Teresa, and the rest of his family was, in a word, incredible. Race fans hung flowers on the fence outside the fourth turn of Daytona International Speedway, near the spot where his familiar GM Goodwrench Chevrolet crashed. In front of Daytona USA, wreaths, photos, and mementos were placed around the fountain where the statue of NASCAR's founder, Bill France Sr., and his wife, Anne, stands. Four-by-six-foot poster boards, covered with hand-written messages of love and grief, were stacked on easels around the makeshift shrine, and grown men wept.

All across the country, the scene was repeated at race tracks, churches, and even supermarket parking lots. A nation of racing fans, diehard and casual, paused to pay their respects to a man they never quite understood, but loved nonetheless.

Dale Earnhardt was many things to many people—hero, villain, friend, and foe—but most of all, he was a legend. *Staff photo*

Harold Hinson

Perhaps the biggest of Dale Earnhardt's 76 career NASCAR Winston Cup victories came in 1998 at Daytona. He ended a 20-year chase for first place in the Daytona 500. *Sam Sharpe*

Dale Earnhardt grew from local racer to regional hero to worldwide legend in his 26-year NASCAR career. *Staff photo*

Dale Earnhardt has always been a fierce competitor. Note the look on his face as he walks away from his wrecked car. *Staff photo*

NASCAR racing has never seen anything like it. Not even the death of the great Glenn "Fireball" Roberts in 1964—described by one sports columnist who knew Roberts well as "awaking to find a mountain suddenly gone"—touched the lives of so many.

The part of the American public that does not follow the sport compared the flood of emotions to the death of Elvis Presley or Lady Diana Spencer. Psychologists were interviewed, seeking answers to the question of how one man could generate such feeling. Those who knew Dale Earnhardt knew the answer. Those who knew him only as a public figure wondered, "Who was Dale Earnhardt?"

Earnhardt was one of, if not the, most talented stock car drivers who ever lived. He raced with a fearless abandon that had even his competitors in awe. Stacy Compton, a rookie driver on NASCAR's elite Winston Cup series last year, put it as simply as it can be put: "He could amaze you on a daily basis." Race fans stand in awe of the skill and daring that all drivers display every time they get into their cars. They live vicariously through them, wishing that somehow, someday, they might be able to drive a car at 200 miles per hour and have the guts to dive into a hole that isn't there. For a driver—and Compton is not alone—to put Earnhardt on a pedestal says something about that skill and daring that perhaps no one else had in such abundance as he.

"I remember at Talladega, as a rookie in 1979, he made one of the most spectacular passes I ever saw," said Jim Hunter, the president of Darlington International Raceway. "They came through the tri-oval at 200 miles per hour and they were four or five abreast, and he happened to be the guy on the bottom.

"As you know, there's only room for like, four in the middle of the tri-oval. Well, he just went on down in the flat part. He had the two left wheels in the grass—at Talladega! I don't know whether he scared the hell out of everybody else, but he did me. It was like one of those kamikaze slides, and he never let out of the gas. He came through there and took the lead.

"And he was a rookie! Everybody said, 'He was lucky; he'll never be able to do that again.' Well, how many times have you ever seen Earnhardt do things with a race car that the more he did it, the more you'd say, 'He can't do that'? But he'd do that. And he just kept doing it."

This is the way most fans will remember Dale Earnhardt—in the cockpit, ready to race, the Intimidator personified. *Nigel Kinrade*

If there was ever a man born to race, it might have been Dale Earnhardt. Born to Ralph and Martha Earnhardt on April 29, 1951, in the small town of Kannapolis, North Carolina, he grew up with a dad who raced and built race cars for a living. In those days, that was something few had the guts to do. He grew up eating, sleeping, drinking, and living racing.

"I can remember him coming to the Columbia [SC] Speedway," Hunter, who was then a reporter for the local newspaper, said. "He was Ralph's gofer. Back in those days they hauled the cars on flatbed trailers. There'd be Ralph and Dale and maybe one pal of Ralph's. Ralph was the driver, the chassis man, everything. But Little Earnhardt used to always be around that car.

"He'd tighten the lug nuts...I can't remember how old he was, but if you go back to the '60s, he'd have to have been 10 or 12 or something—not yet a teenager. He was a quiet little guy.

One thing you never lacked with Dale Earnhardt was action. Here, he's sitting in the middle of plenty on the banks at Richmond. *Nigel Kinrade*

"He was always all over Ralph's car. If Ralph changed a gear, he'd be helping him. He was his daddy's helper. When it came to race cars, Dale was one of those who knew them from top to bottom, side to side, and back to front. He had probably forgotten more about race cars than most of today's drivers will ever know."

There is, however, a great leap between "knowing" cars and driving them. There are hundreds of great mechanics who can't drive a lick that would attest to that. Earnhardt was one of the rare few who was able to make that leap.

"He just had the ability to do things with a race car that no one ever had, and I don't know that anyone ever will," Hunter said. "He had 'feel'—the thing that the truly great athletes have.

"Putting him in the seat of a car was like sending Reggie Jackson up to bat. When Jackson got that bat in his hands, he was going to hit the ball. He was like any great athlete. They just have that ability and that confidence, that attitude that, 'I can do this,' and Earnhardt did it."

Part of Earnhardt's appeal, however, is based on the fact that he wasn't universally loved. In his heyday, everybody cheered for Richard Petty. When Darrell Waltrip was in his prime, no one got boos like "Jaws." From the beginning to the end, Earnhardt prompted the best and worst behavior in race fans.

"Some people would call it an image, but I call it the real Earnhardt," Hunter said. "There was a confidence that made people either happy or mad. His driving style either made people happy or mad. But he elicited emotions. I think Dale relished it. I mean, he enjoyed that—both sides of it. He would say things at times that you knew he was in the real world.

"And he had a knack for knowing when to do what to elicit those emotions. I don't think he spent any time thinking about it or dreaming up things to do in certain situations. He just had an

Behind the gunfighter's façade, Dale Earnhardt was a devoted father, husband, and grandfather. Here, he shares some Victory Lane good times with wife Teresa. *Nigel Kinrade*

innate knack for doing things that just sparked either that loud cheer or that tumultuous boo—from everybody."

If that were the only side of Earnhardt, however, he'd have never become the phenomenon that he did. Despite his lack of formal education, he had an uncanny business sense that made him a millionaire.

He built an image as a competitor who liked to play rough, and parlayed his tough-guy image into a fortune. He went from driving a hand-me-down race car of his dad's to owning three Winston Cup teams of his own.

Kevin Triplett, now NASCAR's director of operations, knew Earnhardt on perhaps more levels than anyone. As a kid growing up in Kingsport, Tennessee, Triplett knew Earnhardt as "One Tough Customer," from the ads of Earnhardt's one-time sponsor, Wrangler Jeans. When Triplett went to work for a newspaper, he knew Earnhardt as a guy who'd give you a straight answer if you asked an honest question.

Later, Triplett said, figuring that he was young enough to go back to newspapering if it didn't work out, he took a job as Earnhardt's public relations manager.

Racing was always a family affair for Earnhardt. Sons Dale Jr. *(left)* and Kerry *(right)* flank their father prior to the Michigan event last year in which all three started. *Nigel Kinrade*

Finally, when he left Earnhardt to go to work for NASCAR, Triplett knew him as the guy on the other side of the fence.

"We always got along, which I never really understood, but we did," Triplett said. "We just hit it off somehow."

At each of those levels, Triplett said, his appreciation grew for a man who was simple on the surface, yet complex. There was a lot more to Earnhardt, he said, than he showed in public.

"He always asked about my wife, my son, Lucas, and he genuinely cared about things that a lot of people who would never think that a guy who could be so gruff could care about," Triplett said. "And I think that's why I'm so appreciative that I got to know him in a way that a lot of people don't.

"I think anybody that worked with him will tell you that they've seen the gruff side and the side that can push and push, because nobody who achieved what he has achieved could do it without having that personality. But I think they'll also tell you there was a side of him that he almost didn't want people to know existed."

A small group of reporters and friends were standing with Earnhardt in the wings before a public function several years ago, and Earnhardt let his guard down, if only briefly. Someone asked him about not so much the demands on his time but on his soul.

Earnhardt won seven titles in Winston Cup, tied with Richard Petty for the most all-time. Here, he celebrates his sixth in 1991. *Staff photo*

"I get letters and phone calls all the time from the parents of kids who are sick, kids who are dying, who want me to call them and just talk to them," Earnhardt said, his face clouding over. "How can I do that? What do you say?"

But, according to Triplett, when he could—particularly if it was a child—Earnhardt would call.

"That's the side of him nobody would ever believe," Triplett said.

"Regardless of how rough his image was, and there were times he did a lot to help that, he cared a lot more than people give him credit for. He did things for people who didn't have what he had and didn't want people to know it, because that wasn't why he did it."

He was also a doting father. Though the first two of his three marriages didn't work out, there is little doubt that he loved his kids. His daughters, Kelley and Taylor Nicole, had him wrapped around their little fingers, but knew the boundaries he set. His sons, Kerry and Dale Jr., knew they had to work for what they got. Kerry worked in his dad's Chevrolet dealership before he began racing, and when

Sweeping up the money was something Dale Earnhardt did whenever possible. Here, he collects some of the million-dollar Winston No Bull bonus for winning at Talladega last year. *Sam Sharpe*

Dale Jr. showed an interest in carrying on the family tradition, he started at square one, working on an old late model before he ever got to drive it.

Whether you were a son, daughter, driver, or employee, you were expected to carry your own weight. Steve Park, one of the drivers who became part of Earnhardt's three-car Winston Cup team, called it "tough love."

"When you're associated with Dale Earnhardt, you don't get a pat on the back for running second," Park said. "You might get a pat on the back for winning a race. It's just the way he's taught us.

"If you want to call it tough love, that's the way I was brought up. If you do something right, you might get one of those half-smiles underneath that mustache, but if you do something wrong, you're out there baling hay in the hot sun with him."

No one may ever completely figure out Dale Earnhardt's mass appeal. Some of those aforementioned psychologists say that it was in part because he was one of us, but one who experienced life to the hilt while most of us nibble around the edges. If we loved him, hated him, or envied him, we could not ignore him.

Someone may replace him on the race track. There are drivers that we perhaps have never even heard of that may match or exceed his skill and daring. But for the total package, from the swagger to the side we almost never saw, Dale Earnhardt evoked a passion in us that we may or may not ever feel again.

"It's going to be tough," Triplett said. "Just like it would have been if something had happened to Richard Petty in Petty's time. I don't think the void will ever be completely filled. Things like that just don't happen.

"But it would be a shame if the things that the drivers learned from Dale, if the things he did for this sport weren't carried on by other people because he did too much for us not to learn from that.

"The void will grow smaller, but it will never disappear."

27

A PAINFUL GOODBYE

Dale Earnhardt Is Gone

BY BRUCE MARTIN
Stock Car Racing May 2001

Dale Earnhardt was a hero, and heroes aren't supposed to die. The man seemed invincible. He was fearless behind the wheel of a race car right up to the last second of his life.

Earnhardt was the definition of a man—strong, rugged, courageous. But he also displayed compassion—at times, deep compassion.

On the morning of the Daytona 500, as he was headed to his black Chevrolet, Earnhardt stopped to hug Kyle Petty, who was returning to Winston Cup racing after finishing last season in the Busch Series. On May 12, 2000, Petty's son Adam died while practicing for a Busch Series race at New Hampshire International Speedway.

Earnhardt hugged Petty for two minutes and told Petty he loved him. Petty cried in Earnhardt's arms. Then, Earnhardt released Petty, patted him on the back, gave that trademark Earnhardt smile, and said, "Let's go get 'em."

Before he climbed into his race car for the last time, Earnhardt hugged and kissed his wife, Teresa. It would be the final hug and kiss of Earnhardt's life.

Earnhardt died in the final turn of the final lap of the biggest race in NASCAR Winston Cup racing—the Daytona 500. His death left a void that may never be replaced.

"It seems like I was put here to drive a race car. I was bred and raised to do that." — Dale Earnhardt. *Sam Sharpe*

Sam Sharpe

Earnhardt's father, Ralph, taught his son the family tradition of tough driving. *Staff photo*

Racing Legend

For over two decades, Earnhardt was one of the sport's greatest heroes. He wasn't just a race car driver; he was a legend.

He had the swagger of Babe Ruth, the attitude of a gunfighter, and the ability to back it up.

Earnhardt's face looked like it belonged on either a Wanted Poster or on Mount Rushmore. He was the epitome of the American race car driver and the biggest name in American racing history to be killed in a race.

Earnhardt always did everything in a big way. His last victory may have been his best, when he drove from 18th place to the lead in the closing laps of the Winston 500 at Talladega Superspeedway last October.

Tragically, his death came in a big way, too.

Two of the cars that he owned, driven by Michael Waltrip and Dale Earnhardt Jr., were in front heading to the checkered flag of the Daytona 500. Earnhardt was in third place, locked in a battle with Sterling Marlin, Rusty Wallace, and Ken Schrader.

At that point, Earnhardt wasn't going to win, but he could help his two cars win the race by running interference. As the battle continued in turn four, Marlin's Dodge touched Earnhardt's Chevrolet, sending it into a slide into the retaining wall.

Earnhardt was dead before Waltrip ever crossed under the checkered flag.

A Giant Void

With Earnhardt's death, NASCAR not only lost its greatest hero—it also lost its soul. Richard Petty may be the King of NASCAR, but Earnhardt was a true hero to the American common folk who worshiped him like a god.

Mike Slade

He was a cultural icon endeared by the masses in the same manner as Elvis. As soon as the news of Earnhardt's death spread, fans flocked to the Dale Earnhardt, Inc., race shop in Mooresville, North Carolina, and created a scene reminiscent of Elvis' Graceland in Memphis, Tennessee.

NASCAR Chairman Bill France Jr. said Earnhardt's death was the most difficult period in NASCAR's 53-year history.

"I can't think of any time that has been tougher," France said. "It's going to take time to fill the void, if we ever fill it. I'm sure we will fill it. Life has to go on. Somebody is going to come along. Curtis Turner was a hard-charging driver in his day...Fireball Roberts, Joe Weatherly. There were a number of drivers in their respective eras. They were the Dale Earnhardt of their day.

"Dale Earnhardt Jr., to me, looks like he has pretty good potential, too, to follow in his father's footsteps."

There lies the greatest tragedy of all in Earnhardt's death. Race fans have lost their hero, but Dale Earnhardt Jr., along with siblings Kerry, Kelley, and Taylor, lost their father.

Common Man

The image of Dale Earnhardt is that of a stock car driver at his best—a man who can seemingly do things with a race car other drivers would fear doing themselves. This persona helped Earnhardt win seven NASCAR Winston Cup titles and gain a mythic presence in the sport.

But there was more to the man than his ability to drive the black, No. 3 Goodwrench Chevrolet Monte Carlo. He was a man who, despite quitting school in the ninth grade, had an amazingly sharp business sense. That helped him amass millions through souvenir sales, a Chevrolet dealership, product endorsements, a poultry farm, and ownership of three NASCAR Winston Cup race teams.

He also was a rugged individualist who was more at home in the outdoors. He loved to hunt. He would prefer sitting on a backhoe, clearing off a few more acres of farm land, over spending his time at a country club playing golf.

MEMORIES

"The thing I remember most about Dale Earnhardt, and the thing that, to me, really epitomizes him, is something that happened at Talladega my rookie season [1994]. The second race there, my team [Billy Hagan] just didn't have a lot of money. We had one good restrictor-plate engine and we broke it in practice. There was no way we were even going to make the race.

"I was walking through the garage before second-round qualifying and ran into Richard Childress. He said, 'How are things going?' and I said, 'Not so good.'

"I told him what was going on but didn't say much more. Even if I had thought to ask him for some help, there is no way we could have afforded anything from them.

"I found out later he told Earnhardt about it and Earnhardt had an idea. He and Richard gave one of their qualifying engines to Dave Marcis, and had Marcis move his qualifying engine to our car. The engine we got was phenomenal. We were 11th-fastest in second-round qualifying and made the race because of Dale Earnhardt and Richard Childress.

"They never asked for anything and they never even told anybody, as far as I know. Dale and Richard helped us out because we needed help. We didn't have the money to pay for it, and they knew that, but they helped us out anyway. Instead of going home, we made the race."

—Driver John Andretti

Earnhardt enjoyed his ability to be a common, everyday man, even though he achieved a level of status that made him one of sports more recognizable figures.

"Around Kannapolis, North Carolina, and Mooresville, North Carolina, I'm just one of the guys," Earnhardt said a few years ago. "It's amazing to think of what somebody from a small town could accomplish."

Once self-conscious when dealing with the media and fans, Earnhardt skillfully developed into one of racing's best spokesmen. Once, he was asked whether he considered himself stock car racing's version of Babe Ruth.

"Babe Ruth?" Earnhardt asked incredulously. "What about Richard Petty?

"I feel honored to be one of the greats. I go out and race hard and try to focus on what I do and try to be the best I can be. Sometimes, it doesn't come. Sometimes, it does; but I have been fortunate enough to be with a great race team that provides that opportunity for a driver to go out there and do as well as I have. It has been a team effort.

"It seems like I was put here to drive a race car. I was bred and raised to do that."

Race fans who witnessed some of the daring moves Earnhardt has made on the race track will one day tell their grandchildren of the time they were there to watch Earnhardt and the famed "pass in the grass" at the Winston in 1987. Others will remember Earnhardt at Bristol in 1994 splitting between two cars in a hole that looked narrower than the width of his Chevrolet.

Earnhardt's aggressive style earned him the nickname the Intimidator. It also earned him a legion of fans unlike any other in auto racing history.

From his earlier days in Winston Cup, such as this 1980 race at Martinsville (*top*), to later in his career, such as the 1995 running of The Winston, Earnhardt showed no fear in swapping paint with the best of them. *Staff photo/Sam Sharpe*

As Earnhardt's popularity soared, so did his business interests. He wouldn't cut a driver any slack on the racetrack, but he cut many major deals away from the speedway. *Harold Hinson*

Born to Race

Earnhardt accomplished much in his life, which ended just shy of his 50th birthday. At one time, however, it appeared he had made a decision that would leave him behind.

When Earnhardt was 16, he quit high school so he could work on his father's race car. Ralph Earnhardt was a legendary driver in NASCAR's Late Model Sportsman division, which has evolved into the NASCAR Busch Series. He believed his son had made a terrible mistake.

"I didn't like school, I wanted to be home working on Dad's race cars," Earnhardt once said. "I didn't like school, I wanted to be home cleaning up the shop. I didn't like school, I would just as soon be washing wrenches."

Earnhardt persevered and eventually earned his father's respect.

"The Man in Black" took his car to the front of the 2001 Daytona 500, and later died trying to keep competitors from reaching Michael Waltrip. *Sam Sharpe*

Ralph Earnhardt died of a heart attack in 1973 while working on his race car in Kannapolis. Dale found his father slumped over the car.

"It's tough to lose people that you love—like my father and my uncle who I used to hunt and fish with over the years," Earnhardt said. "For the first two years of my dad's death, I stayed pissed off at the world—not in the way everybody knew it, but inside. To finally learn to live with the fact my dad was gone or wasn't there to help or make the right decisions or ask what would he do, I took good memories of him and went on."

Little did Dale Earnhardt realize when he quit school that he would become a diversified, multimillionaire businessman who would have to worry about trademark and copyright laws, rather than setting up his stock cars to win races.

"It is really awesome for me when I look at the TV and there is my son in a commercial," his mother, Martha Earnhardt, once said. "Everywhere you look, there he is. I go into the grocery store, and there he is on a cereal box. It is hard to comprehend how far he's come."

Earnhardt sits in his car awaiting the start of this year's Daytona 500. Little did the world know that this would be his last race. *Sam Sharpe*

Growing Up

The eldest son in the family, Earnhardt was just one of the boys when he was growing up with his two sisters, Kathy and Kaye, and two brothers, Danny and Randy, in the North Carolina textile mill community located 25 miles northeast of Charlotte.

"The kind of child Dale was with me, him being my first son, I think his dad accused me of spoiling him a little bit," Martha said.

To the man who is now a hero to millions of race fans, Dale Earnhardt did things regular kids did like play baseball, football, and basketball, but the things that really intrigued him involved cars.

Young Dale idolized his father, and was always eager to help his dad in his race shop, which was in the garage next to the house on Sedan Avenue in Kannapolis. It was an easygoing lifestyle in the late 1950s through mid-1970s.

"When I didn't have to get up and go to school, I would hang around the neighborhood with the guys down at the creek or play in the woods," Earnhardt said. "We would race bicycles and work on go-karts. I was always tearing something apart and trying to put it back together again. I was always tinkering."

MEMORIES

I knew a pool hustler. I was part of his game, although I can't remember making more than one shot on a nine ball and it was straight in, 2 inches from the pocket.

My part in the great hustle consisted of being Dale Earnhardt's partner. Earnhardt was good. He could make the cue ball shell butter beans and play the piano. In pool, as in racing, he played to win. Winning at everything he undertook was a trait of his character.

If people were present, he would announce: "Me and Phillips will play any two of you for $20 a game."

The first time I heard this, I whispered while he was chalking his cue stick: "I'm not good at shooting pool, partner. Why don't we just play ice hockey or go sky diving?"

"You don't have to be good at pool," he snapped. "I can whip any two of 'em here. I'm using you as a decoy. It sounds better than me saying, 'I'll take on any two of you by myself.'"

—Stock Car Racing *columnist Benny Phillips*

Earnhardt remembered those youthful days at the race track as his father prepared to do battle in the short track "bullrings" of the Southeast.

"When we first got to the race track, all the guys would unload their cars," Earnhardt said. "Before they started practice, they would let me stay in the infield. I would hang around Tiny Lund's car and look at everybody's different cars. Guys like Little Bud Moore and Marion Cox. He had a bunch of kids, too. We played and fought and argued with them. We would do all kinds of junk like that.

"I would ride in the water truck with the guy who watered the race track in the evenings if I got there early enough. It was just being a kid."

And Earnhardt knew how to have fun at home, too. He used to put model race cars together and was an avid slot-car racer. He also learned how to build bicycles and go-karts.

"After supper, all the neighborhood kids would get together and you would time each other around the block to see who could ride around the block faster on a bicycle," Earnhardt said. "It was always racing; it was always competitive stuff like that. Or who could run the fastest, the furthest.

"We played ball, and we played football all the time around the neighborhood, but we were always racing something, too."

Martha Earnhardt would later take exception to Earnhardt's image as an intimidating race driver—"the Man in Black."

"I think they have carried that too far," Martha said. "It makes me mad sometimes the way they portray him. It hurts me as a parent. I don't care how old your child gets, he is still your child. Any time anybody says anything bad about him or derogatory about him, it hurts me.

"I've tried to make myself get used to it, but it is still hard to do."

Just Drive

But despite the luxuries that come with the money and the adulation, Dale Earnhardt longed for those simpler days of his childhood in Kannapolis.

Fan tributes outside of Dale Earnhardt Inc. *Sam Sharpe*

Sam Sharpe

"Dale told me one day, 'Mom, I wish I didn't have to do anything but get in that race car and drive it,'" Martha said. "I told him, 'Son, you shouldn't have been so good at it, you wouldn't have had to worry about it.'

"He just wants to drive a race car. That is all he wanted to do in his life—drive a race car."

And that is what Dale Earnhardt did all the way up to the last moment of his life.

"It seems like that is all I ever knew or wanted to do was drive a race car," Earnhardt once said. "I don't think God said it as much as I was born into that life.

"I was born into racing; I didn't just decide or happen to be there."

Earnhardt enjoyed the success and accolades for a lifetime of hard work. He had more racing talent than perhaps any driver in NASCAR history.

And then, in an instant, he was gone.

Reactions: What People Had to Say About the Death of Dale Earnhardt

"On behalf of Taylor, Dale Jr., Kerry, Kelley, and the entire Earnhardt family, which also includes all the employees of Dale Earnhardt, Inc., we want to sincerely thank everyone that has offered their condolences during our very tragic loss. This outpouring has been overwhelming. We kindly ask that you keep our family in your thoughts and prayers."
—Teresa Earnhardt

"We're deeply saddened by the loss. We appreciate everybody for their support. We'll get through this. I'm sure he'd want us to go on, so that's what we're going to do."
—Dale Earnhardt Jr.

Sam Sharpe

"NASCAR has lost its greatest driver, and I personally have lost a great friend."
—*NASCAR Chairman Bill France*

"Dale Earnhardt was much more than a race car driver. He was a very loving husband and a proud father and grandfather. He was a successful businessman. He was also a hero to millions of racing fans throughout the world. Dale was my friend. We hunted and raced together. We laughed and cried together. We were able to work side by side and have the success we had for almost 20 years because we were friends first. I will miss him always. He was the greatest."
—*Car owner Richard Childress*

"After the race was over, I heard things didn't look very good but, man, Earnhardt. You figure he'll bounce right back. Your first thought is, 'Hey, he'll probably come back next week at Rockingham and beat us all.'"
—*Driver Jeremy Mayfield*

"Like all of the NASCAR family, I was stunned and saddened by the loss of Dale Earnhardt. We shared a common bond in championships as well as a mutual respect. Our family has raced against his family since this sport began, going back to when my dad and I raced against Ralph Earnhardt. My thoughts and prayers and that of the entire Petty organization are with Teresa, Kerry, Kelley, Dale Jr., Taylor, and the rest of the Earnhardt family."
—*Richard Petty*

MEMORIES

"I have so many Dale Earnhardt stories I could write a book. I remember he came to my mom and dad's house in Berwick [Pennsylvania] and I'll never forget it. We crashed three or four cars one night in my dad's junkyard and just had a great time afterwards just laughing and eating dinner.

"Then there was the time in 1988 when we raced up Route 136 from his old shop in Kannapolis to his new facility. It was late at night. He had a pickup and I had a pickup and he beat me. He never let me forget that and rubbed it in every chance he got."

—Driver Jimmy Spencer

"When he was at the top of his game, he was amazing. He could do things with a race car that you didn't think anybody could do. There was a time when you could see a 20-car pileup and, if just one car made it through, it was the one Earnhardt was driving. For a lot of fans, Dale Earnhardt was what they thought about when they thought about NASCAR racing. He could do so much and was so talented. He knew it, and he knew you knew it."

—Driver Kyle Petty

"Dale Earnhardt was a winner, through and through. He won when he started on the short tracks, he won in Busch, and he won in Winston Cup. I can't remember a Winston Cup race where, when you made a list of who you had to beat to win, that list didn't include Dale Earnhardt."

—Driver Joe Nemechek

"We shared many things in our professional life, as well as our personal life, with Dale. We shared the victories in racing and the victories in championships. I thanked him for those victories and I thanked him for helping me climb to where I am today. My regret is that I never thanked him for teaching me the most important lesson in life and I never told him how much I love him. I hope and pray that he understood."

—Car owner Andy Petree

"We always thought of Dale as being invincible, so when he didn't climb out of that car after the wreck I knew it was bad. I talked to him this morning before the race and he was poised to challenge for an unprecedented eighth championship and was really looking forward to a fantastic season."

—H.A. "Humpy" Wheeler, president and general manager of Lowe's Motor Speedway

"It's hard to put into words what the loss of Dale Earnhardt means. I feel without question he was the greatest driver we have ever had. At a time when the popularity of NASCAR has been rapidly growing across this country, we can thank Dale Earnhardt for a lot of that attention. The fan base he enjoyed is unparalleled in this sport."

—Clay Campbell, president of Martinsville Speedway

Sam Sharpe

"There are no words to express the extreme sense of loss we are feeling now. Dale's greatness as a champion speaks for itself. The significant loss is of the man himself; we've lost a dear friend."
—Jay Signore, president of the IROC Racing Series

"He had his own way of doing the best he could on the track. We all may not have approved of the way he did things on the track, but he did win seven championships. It may not have been the best way, or the prettiest way, but he always got the job done."
—Race fans Dana and Melisa Williams

"Brooke [Gordon] and I are deeply saddened by this devastating loss. Not only is it a huge loss for this sport, but a huge loss for me personally. Dale taught me so much and became a great friend."
—Driver Jeff Gordon

MEMORIES

Earnhardt went to Alabama to go deer hunting with friend Neil Bonnett in the early 1970s, just when three-wheelers were coming on the market.

Bonnett had purchased a new bike. Earnhardt arrived Sunday night, looked over Bonnett's new toy, and had to have one immediately. Bonnett called the dealer, got him out of bed, and explained that his buddy Earnhardt had to have a new three-wheeler.

So be it.

Next morning the two go hunting. About the middle of the morning, they start out of the woods.

"All of a sudden I get this Wham! Wham! Wham! on the left rear of my ATV," Bonnett said. "Earnhardt has run into me. I stop, get off, and look at the damage. The fender is scratched and cut in one place.

"I turn around, hands on my hips, and say to him: 'Why the hell did you do that?'

"He gives me that sly smile, then replies: 'It's time to wear the new off, partner. Now that you have a scratch on the fender, you won't have to be so careful.'"

—Stock Car Racing *columnist Benny Phillips*

"I will tell you that some of the fiercest and most successful drivers are also the most aggravating on the track. Dale was incredibly tenacious; he drove me to rise to his level. He made me want to be the best, because he made me want to beat him. I have never in all of my experience raced against anyone with as much desire to win as he had and that's saying a lot, because I've raced against them all."
—Driver Mark Martin

"He's tough as hell on the track. He never gave anything, but you always knew what to expect, and he had a lot of respect for it, because he treated everyone equally. He was really frustrating to drive against, because there were times when he could have made it much easier, but that wasn't in his personality. You always had to work that much harder with him."
—Driver Jeff Burton

"I was in total shock. I've seen him hit, and flip, and tumble, and me and him have been in some at Talladega—upside down and I couldn't believe it. It made you just want to go throw up, just sick at your stomach. You couldn't believe it could happen."
—Driver Sterling Marlin

"I, like everyone else, am in shock with the passing of Dale Earnhardt. Besides being an incredible driver and spokesman for the sport he so loved, he was a true friend and has been a major influence on my life and career."
—Driver Bobby Labonte

"Remembering back to '79, I went to my first NASCAR race at Riverside Raceway. There I saw a guy in a No. 2 car. As we all know it was Dale Earnhardt. I said to myself, 'He's the man.' From that day on, I've been a fan of 'the Intimidator.' If you loved him he was your hero. If you hated him, he was your biggest fear on the track."
—Race fan Bill Bennett, Auburn, CA

"Dale Earnhardt made a difference in the world. On the track, he made us all better drivers because he set a standard of excellence we all aspired to achieve. He had a passion and a desire that took the sport of NASCAR to a new level every time he climbed in the car. Off the track he was a kind, giving, loving man who gave his all to his family and friends. He worked tirelessly to make the world a better place for as many people as possible who were less fortunate than he."
—Driver Tony Stewart

"Just 30 days ago I lost my father to cancer. Not only was he my father, best friend, and counselor, he was my hero. Dale Earnhardt, for me, will be the other hero I've lost this year. My heart and prayers go out to everybody that was associated with Dale. As my father will always be remembered in my heart, so will Dale."
—Race fan David Taylor, Mobile, AL

DALE EARNHARDT CAREER STATISTICS (1975–2000)

Year	Car Owner	Races	Wins	Top 5	Top 10	Bud Poles	Winnings	Point Standings
2000	R. Childress	34	2	13	24	0	$4,918,886	2nd
1999	R. Childress	34	3	7	21	0	$3,048,236	7th
1998	R. Childress	33	1	5	13	0	$2,990,749	8th
1997	R. Childress	32	0	7	16	0	$2,151,909	5th
1996	R. Childress	31	2	13	17	2	$2,285,926	4th
1995	R. Childress	31	5	19	23	3	$3,154,241	2nd
1994	R. Childress	31	4	20	25	2	$3,300,733	1st
1993	R. Childress	30	6	17	21	2	$3,353,789	1st
1992	R. Childress	29	1	6	15	1	$915,463	12th
1991	R. Childress	29	4	14	21	0	$2,396,685	1st
1990	R. Childress	29	9	18	23	4	$3,083,056	1st
1989	R. Childress	29	5	14	19	0	$1,435,730	2nd
1988	R. Childress	29	3	13	19	0	$1,214,089	3rd
1987	R. Childress	29	11	21	24	1	$2,099,243	1st
1986	R. Childress	29	5	16	23	1	$1,783,880	1st
1985	R. Childress	28	4	10	16	1	$546,596	8th
1984	R. Childress	30	2	12	22	0	$616,788	4th
1983	B. Moore	30	2	9	14	1	$446,272	8th
1982	B. Moore	30	1	7	12	1	$375,325	12th
1981	R. Osterlund J. Stacy R. Childress	31	0	9	17	0	$347,113	7th
1980	R. Osterlund	31	5	19	24	0	$588,926	1st
1979	R. Osterlund	27	1	11	17	4	$264,086	7th
1978	W. Cronkrite R. Osterlund	5	0	1	2	0	$20,145	43rd
1977	H. Gray	1	0	0	0	0	$1,375	—
1976	W. Ballard J. Ray	2	0	0	0	0	$3,085	—
1975	E. Negre	1	0	0	0	0	$1,925	—

DALE EARNHARDT CAREER HIGHLIGHTS

1975—Entered first Winston Cup race. Started 33rd and finished 22nd, one position ahead of future car owner Richard Childress.

1979—Winston Cup Rookie of the Year. Won first pole.

1980—Won first Winston Cup championship in only his second full season.

1983—Won a 125-mile qualifier for the Daytona 500, beginning a long string of successes in the 500 qualifying races.

1986—Won second Winston Cup championship.

1987—Won third Winston Cup title. Became third modern-era driver to win four straight races.

1990—Won fourth Winston Cup championship. Won first of four IROC titles.

1991—Won fifth Winston Cup championship.

1993—Won sixth Winston Cup championship.

1994—Won seventh Winston Cup championship.

1995—Won sixth Bud Shootout.

1996—Won pole at Watkins Glen with a track record two weeks after breaking his collar-bone and sternum in a crash at Talladega.

1997—First driver to reach $30 million in all-time American motorsports winnings.

1998—Won first Daytona 500.

1999—Won 10th consecutive Twin-125 qualifying race at Daytona.

2000—Finished second in Winston Cup points. Climbed from 18th place over the last five laps to win the fall race at Talladega.

Staff photo

Sam Sharpe

Sam Sharpe

Sam Sharpe

KING OF BEERS
MONTE CARLO
MOOG
CLEVITE
SIMPSON
Edelbrock
WIX
plasti-kote
76

DALE EARNHARDT JR.

STANDING TALL IN THE SHADOW OF A LEGEND

LARRY COTHREN

FOREWORD BY
BENNY PHILLIPS

Lowe & B. Hould
Publishers

FOREWORD

BY BENNY PHILLIPS

On a May night that should have come rushing in like Christmas morning, Dale Earnhardt was madder than a pit bull at a poodle show. Dale Earnhardt Jr. had just captured the pole position for the Coca-Cola 600 at Lowe's Motor Speedway. The media rushed to the seven-time Winston Cup champion's truck to get "proud pop's" reaction.

Quicker than a jack-in-the-box, it was easy to see that Earnhardt Sr. was in a sullen mood. "Why don't you go ask him about the pole?" he said, "Hell, I wasn't driving the car, he was."

So members of the press made their way to see the kid who had just pulled off another major surprise. Grinning (and trying not to) like a possum eating persimmons, the kid explained how he drove the track—high into this turn, low into the next, and keeping a tight line all the way. He charmed the press like a grizzled old veteran.

The year was 2000 and "Little E" was a rookie on the Winston Cup circuit. He rocked the racing world with victories at Texas and Richmond, then booted experienced drivers aside and won The Winston at Charlotte during May.

He capped that impressive run by winning the pole for the 600 at Charlotte. What father would not be glowing like a Halloween lantern? Dale Earnhardt Sr., that's who.

"Why are you so ill?" I asked him the next day as he stood behind his rig.

"Who said I was ill," he snapped.

"Are you saying you were not ill last night?" I asked.

He smiled that little sly smile.

"Are you not proud of your own son?"

"You know I am," he replied, "and maybe I'm expecting too much from him. I wasn't mad about what happened last night. I guess I was still upset about what happened a couple of nights ago.

"It was after midnight, and I went over to his house and broke up this party. There were a bunch of his friends there. Listen, he can have parties. I like parties, but there is a time for everything, and I am so concerned about him being totally focused on racing. He should have his mind on the race this week instead of throwing a party. I think my greatest fear is that he will not be totally focused on racing.

"My son is not going to run a boot barn one week, operate a car dealership the next, play a guitar in between time, and ride a motorcycle around the country in his spare time. No! He's not going to do that and expect me to help support his racing efforts.

"I want him totally focused on racing. I'll be happy with that."

THE BEGINNING

1

From Cut-Out Cars to Myrtle Beach

BY LARRY COTHREN

The first race car Dale Earnhardt Jr. remembers putting his hands on was made of paper. He was just a toddler at the time, and he had only a vague notion of his father who was struggling to make it as a stock car racer.

"I didn't have any Matchbox cars, so I cut pictures of cars out of magazines and raced them at a table, as you would Matchbox cars," recalls Earnhardt Jr. His parents were divorced by the mid-1970s and the young tabletop racer was living with his mother.

"I knew a little bit about my father, but not a whole lot," he says. "I had never watched racing or known about racing at that time. I think I was interested in it at a very early age, even given the circumstances. Then once I did go live with my father, it just became more of an obsession. There was nothing like being eight, nine years old and playing with Matchbox cars in the living room while listening to Barney Hall talk about the race on the radio. It was just a great time, and it was fun. That escalated into going to the races as I got older."

Even then, Earnhardt Jr. had a sense of his rightful place in the sport.

"I wasn't really into working on the cars or knowing or learning about the cars," he says. "I was just really interested in what it was like being a driver and the things you had to do. I started watching my dad more, as far as watching him drive and trying to figure out why he did this or why he did that, or why the

By age 18, Earnhardt Jr. had become a full-time racer. *Harold Hinson*

Dad hangs on to "Little E" at Charlotte Motor Speedway in the mid-1970s. Others pictured are Dale Sr.'s brothers, Randy (middle) and Danny, and Danny's wife Sherry. *Gary Hargett Collection*

car wasn't working and what he was doing to fix it and how they fixed it.

"Then when I turned 13, I remember sitting there one night in the living room with my dad and I wanted so badly to race, but I knew it was three more years before I even got a driver's license. I just could not comprehend having to wait three more years. I was about to go crazy. You know what I mean? That's how bad I really wanted it at that point. I told Daddy, 'I've got to do something because I can't wait for three more years.' It was just too long. I had never wanted something or craved anything as bad ever before. He just told me, 'Man, you've got to be patient.' I don't think he took me quite as seriously as I was at that time."

Earnhardt Jr. is shown with his late-model Chevrolet at North Wilkesboro Speedway in 1993. *Harold Hinson*

No Mercy

The skinny teenager who showed up at Andy Hillenburg's driving school in 1992 was much like hundreds of others who have graduated from the program. Young racers chasing a dream and looking for an advantage come through Hillenburg's school week after week, month after month, year after year.

This kid was different though. The shy youngster with the slight build showed up as the son of the greatest stock car racer ever. He was the third generation in a line of stock car thoroughbreds who epitomized hard-core, stand-on-the-gas driving. From the very moment Dale Earnhardt Jr. decided to become a racer, he was sure to have that fact hovering over him. That was true when he showed up at Hillenburg's driving school as a 17-year-old wannabe, just a month away from his 18th birthday, and it would be true for the rest of his life. Being a stock-car-driving Earnhardt would mean more than being just another kid in a driving school.

Dale Earnhardt Jr., then just a wisp of a teenage boy, quickly lived up to the Earnhardt name during that week at Hillenburg's school in September of 1992. When the engine exploded in one of Hillenburg's cars with Earnhardt Jr. behind the wheel, it blew a connecting rod clean out of the block. Granted, the car did have an experimental engine because Hillenburg's mechanics were searching for the right engine combination to use in the school's cars. But Hillenburg says, "He didn't show it any mercy."

Yes, this kid was an Earnhardt.

"You could tell even then that he was determined to try to make it work," says Hillenburg, "and you could tell that if he didn't have the experience, he was going to get it and if he didn't have the talent, he was going to figure it out and get the talent. There was a lot of determination on his part."

Hard driving and determination are dominant traits in the Earnhardt gene pool. Ralph Earnhardt, Dale's father, was a terror on dirt tracks around the Carolinas in the 1950s until his 1973 death from a heart attack at age 45. Long before Dale became The Intimidator, Ralph was tearing

At age 17, Earnhardt Jr. attended Andy Hillenburg's driving school at Charlotte Motor Speedway. *Stock Car Racing Archives*

it up on short tracks, and he became the standard by which Saturday-night racers all around were measured. That was true at Concord, Charlotte, Hickory, Columbia, Gaffney, Asheville, Greenville, Spartanburg, Lancaster, Monroe, and every other short track around the two Carolinas. "Anywhere he went, when he pulled in you knew he was the man to beat," recalls Ned Jarrett, who competed against Ralph in NASCAR's Sportsman division, primarily in the 1950s when both men won national titles.

When Dale began eyeing a career as a stock car driver in the late 1960s and early 1970s, he had to make his own way. He learned what he could from Ralph, but Dale built his own cars and put together his own deals to drive the cars of others. Ralph made sure that Dale learned the trade on his own, and that became the Earnhardt way.

Twenty years later when Junior began embarking on a driving career, he was expected to work on his own cars and learn lessons the hard way. "I don't know that Dale ever really sat down and actually told Junior how to drive a race car," says Gary Hargett, who owned the late-model Chevrolets that Earnhardt Jr. drove in the mid-1990s. "He never did. He never would go to the race track with him, and I think a lot of that came from the way Ralph was. Dale wanted Junior to know if he was going to learn, he was going to learn it on his own."

So while Earnhardt Jr. no doubt enjoyed advantages that other young drivers could only dream about, he was essentially left to his own devices. That, and the pure grit he soon displayed, helped Earnhardt Jr. earn respect from his fellow competitors, such as those he raced against at Myrtle Beach (South Carolina) Speedway, where he and Hargett raced regularly in the late-model division.

There were, nonetheless, some hard lessons to be learned. Hard-core racers populate Myrtle Beach Speedway on a Saturday night. Some are racing just as a hobby, but many hope to be the next Dale Earnhardt, Rusty Wallace, or Jeff Gordon. These are guys who would take out *anybody* for a shot at victory lane. If they happened to take out Dale Earnhardt's kid, then all the better. Plus, it was hard for Junior to shake the "rich kid gone racing" perception that many of the competitors had of him at the fast little half-mile track. It all combined to make Myrtle Beach Speedway a tough place to send an 18-year-old to learn the ropes.

Hargett, a short-track veteran now in his fifth decade as a car owner, teamed with Junior for a few races in 1992, the year Junior turned 18 years old. Then Hargett and Junior ran the full late-model schedule at Myrtle Beach Speedway for three consecutive seasons from 1993 to 1995.

"The first year we ran down there everybody wanted to turn Junior around," says Hargett. "I mean, it would be every week. I told him he was just going to have to take it because they were all going to do it. I said, 'Everybody who races wants to run over Dale Earnhardt. Well, they're never going to get the chance to run over your daddy, so they're going to take it out on you.' And they roughed him up bad that first year."

Before the last race that season, Hargett put the word out that he would no longer tolerate other drivers running into Earnhardt Jr.

"I kept Junior off of you all year," Hargett said at the drivers' meeting. "You've had your fun, you've run over an Earnhardt all year long, and it's over with now. For the next man that hits him, I'm going to let him hit you back. You'll just keep doing it if you want to."

Although just a teenager, Earnhardt Jr.'s driving style proved to be true to the family name at Andy Hillenburg's driving school. *Stock Car Racing Archives*

The message didn't sink in until late in the race when Earnhardt Jr. was on his way to winning.

"There was a boy down there from Monck's Corner [a small town in South Carolina] who turned Junior on the last lap, and we were leading the race," says Hargett. "He spun him out, and I told Junior, 'The next race you run, you need to get him right quick. If you don't, they're going to keep doing this to you as long as you race.'

"We went back to a big race they have down there, a special event in November, and I had really forgotten about it. But we started, I don't know, 9th, 10th, 11th, somewhere along there, and this boy was on the inside of the row right in front of Dale Jr. Well, they gave the green, and when they went down into turn one, Junior just turned him the wrong way and he went through a steel gate down there. Before they got to the backstretch, Junior said, 'Well, was that good enough?' I

The cockpit of a race car would become a familiar place.
Stock Car Racing Archives

said, 'Yeah, if you didn't kill him, it is.'

"Every week, they would beat on that back bumper, beat on that back bumper. But I don't know of anybody else who every really turned him after that."

It wasn't the first time Junior had taken the initiative. Earlier that season, a squabble with another driver ended on pit road as Junior confronted the guy before the start of a race, leaving Hargett with one of his favorite stories. "Dale Jr. stopped him," recalls Hargett, "and told him, 'Look, we can solve this problem we've got now. I'll put my race car out on the race track and you bring yours out there. We'll take off and you go that way and I'll go this way and we'll run head-on on the backstretch. The boy said, 'You're crazy.' And Junior told him, 'That's exactly right—I'm crazy.' And they never had any more problems."

"Well, talk to the man."

To get to Gary Hargett's garage, drive about an hour southeast of Charlotte and its deeply rooted stock car culture, where most of NASCAR's top teams are located. Go through Marshville, North Carolina, home place of country music star Randy Travis, and turn off of Highway 74 to cover the last eight miles on a two-lane road that stretches across gently rolling North Carolina farmland. Go a mile or so past Hargett's Store at the crossroads and turn right onto a gravel driveway beside the wooden roadside sign etched with the words "Gary Hargett Racing." The simplicity of the surrounding countryside and the small garage sitting beside the short gravel road is deceiving, for there's nothing simple or nondescript about Gary Hargett's 40-plus years of

Earnhardt Jr. raced three full seasons at Myrtle Beach Speedway in the mid-1990s. *Gary Hargett Collection*

making a living as a racer.

Hargett teamed with driver Harry Gant in the early 1970s and won Sportsman (now NASCAR Busch Series) races all over the country. Gant and Hargett won 38 of 43 races at one point in the mid-1970s, before Gant moved on to a successful career in NASCAR Winston Cup. Hargett later fielded a Sportsman car for Dale Earnhardt Sr. and won with him. When he was getting his start in the 1950s and 1960s, Hargett rubbed elbows with the likes of NASCAR legends Ralph Earnhardt, Curtis Turner, Speedy Thompson, and Jack Ingram.

Hargett's small office, tucked inside a corner of the garage, is filled with racing photos and memorabilia collected during his five decades in the sport. There's a framed picture of Ralph Earnhardt sitting on the back of his Camaro sometime around 1970. Another picture shows Dale, shirttail hanging out of his jeans, standing outside Hargett's garage beside a Camaro he and Hargett campaigned when they first started racing together in the mid-1970s.

Hargett was tight with the Earnhardts. He bought a new race car from Ralph in the early 1960s and Ralph even occasionally raced the car for Hargett. Later, Dale and Hargett formed a

Earnhardt Jr.'s sister, Kelley, looked on during an interview at Myrtle Beach Speedway. *Gary Hargett Collection*

The number 3 late-model Chevrolet gets serviced at a short track near Greenville, North Carolina. *Gary Hargett Collection*

Junior accepts an award for finishing fourth in late-model stocks at Myrtle Beach Speedway in 1995. *Gary Hargett Collection*

friendship, raced together, and then kept in touch while Dale was becoming a stock car icon in the 1980s. During that time, Hargett was working on a Winston Cup car for Mike Potter and he and Dale occasionally talked at the race track. Hargett is quick to point out that the relationship was hot and cold; it was built upon and held back by the tempestuous nature of both men.

Although out of Winston Cup by the late 1980s, Hargett continued to race local tracks and field cars for whoever was willing to pay the bills. In the 1990s, Hargett began crossing paths again with Dale Earnhardt. Gary Hargett and the Earnhardts began an association that lasted until Earnhardt Jr. moved on to run his own late-model deal for a season or two, and then moved full-time to the Busch Series.

It was 1991 when Hargett, while fielding a car for an aspiring teenage driver at Concord (North Carolina) Motorsport Park, first took notice of Earnhardt's kid.

"Dale Junior came out to Concord and had a little street stock he ran with [his half brother] Kerry," recalls Hargett. "They would swap back and forth. I was at the race track one day and I told Dale, 'How about letting me put Junior in a late-model car?' He said, 'Ahhh, you don't want to fool

Earnhardt Jr. in action at Myrtle Beach Speedway.
Kevin Thorne

with him.' That was just the way Dale was. Every week I would ask him again and we would talk about it.

"The funniest thing is he asked me one day, 'You know which one you're talking about now?' Because at that age, Kerry looked just like Dale Sr., and Dale Jr. looked like his mother. He didn't look like an Earnhardt. Junior looked like he took from his mother and [her father] Robert Gee's side of the family. He had almost blonde hair, and he was a little bitty fellow. Kerry was tall, had dark hair, and a moustache. Dale asked me, 'You sure you know which one you're talking about now?' I said, 'Yeah, I know exactly which one I want.' Junior was just as wild as wild could be, and he would run wide open all the way around the race track in that street stock, just smoking the tires.

"Well, Dale finally broke down. I think what really did it was we were over at Charlotte practicing and I told Dale about it again. He said, 'Man, I don't want to fool with that. You don't want to do that.' Well, I got mad and just turned around and walked off. I was going down through the parking lot and some man hollered at me. I turned around and it was [Richard] Childress [who owned Dale Sr.'s number 3 Chevrolet]. Richard came up there and said, 'You not getting anywhere with him.' I said, 'No, that man is not going to spend any money.' He said, 'Well, if he won't help you then let me know. I'll help you out. We'll put him in a car.'

"Anyway, a day or two after that, Dale called me. I knew they had talked. Dale told me to come over to his shop and talk about this thing. I went over there to Mooresville—where they're at now, but they just had that one old shop then—and I went in and he said, 'You still want to do that deal?' And I said, 'Yeah.' Dale said, 'Well, let's go out here and talk to him.'

Car owner Gary Hargett (far left) joined Earnhardt Jr. for a personal appearance at Myrtle Beach, one of the first such appearances by Little E. *Gary Hargett Collection*

"Junior was out there sanding on a car in a paint booth. He was a little bitty fellow. He didn't look like he was 12 years old, but I guess he was 16 or 17 at the time. Dale Jr., I really believe, was always afraid of Dale Sr. He wouldn't say a word around him. He wouldn't even look at him eye to eye. It was like he was always scared to death. We went back to where Dale Jr. was working and Dale said, 'Junior! Do you want to drive this man's race car?' Junior just glanced up and said, 'Yeah, yeah,' and kept on working. He wouldn't even look at me. Dale said, 'Well, talk to the man then if you want to drive his race car!'

"Dale left and Junior was just like a little kid then. He was ready to go racing. So we worked out a deal. I had a car I had built for the other boy who was driving. It was a true late-model stock. We worked out a deal where Dale [Sr.] paid for the motors and the tires, and we just went out, put him in it, went to Myrtle Beach, and went at it."

The goals were simple, according to those associated with the late-model effort. Ernie Hammonds became acquainted with Earnhardt Jr. and Hargett soon after they began racing at Myrtle Beach. The three formed a bond, with Hammonds eventually serving as parts courier during the week, chauffeur for Junior on weekends, and crewman on race day. "When I would go to Mooresville, to DEI," recalls Hammonds, "and pick up the oil and whatever parts we needed for the race car, Dale would say, 'Seat time and laps is all I'm interested in. Then when we get to the Busch Series, we'll worry about winning races.' And evidently he knew exactly what he was talking about."

Rednecks from Mooresville

BY LARRY COTHREN

Although most of Dale Earnhardt Jr.'s early short-track experience came at Myrtle Beach Speedway, it wasn't the only track where he and car owner Gary Hargett raced in the mid-1990s.

While they were well accepted at most places, they sometimes visited tracks where the locals were less than thrilled to see an Earnhardt. That was the case on their first trip to a short track near Greenville, North Carolina, where the owner offered appearance money for them to show up and race.

"That was Rusty Wallace country," says Hargett. "When I got there I found that out right quick. We go out there and earn the pole, set a new track record the first time we went. In the race, a guy blocked us in on pit road and wouldn't let us out, so we lost a lap. We came back out, got our lap back, and finished second or third maybe, but we had to fight our way out of the race track.

"I was up on top of the trailer and those people wouldn't let me down. Junior drove the car up in the trailer and we shut the door and left him in there, got in the truck, and drove out of the race track. We got out on the road and the man came out there and gave us our money and asked us if we would come back the next week. I said, 'No, I don't think so.' But we went back three or four times and always ran good there. Those people did not like Earnhardts in that part of the country, but that promoter loved it."

Sheet metal hanging in Gary Hargett's garage commemorates a late-model race at Hickory, North Carolina. *Larry Cothren*

Once at a race at Hickory (North Carolina) Motor Speedway, the crowd—and Earnhardt Jr.'s crew—became unruly because of an incident during the race.

"We were like 3rd or 4th fastest in practice but we were 35th or something in qualifying," recalls Hargett. "So they came up with a track provisional to put Junior in the race. Anyway, we started last, and we were running over everybody on the race track and got to third or fourth. Then they had a red flag for a funeral, which they do at Hickory. If they have a funeral over there, they stop the race because the cemetery is right beside the race track. So they stopped the race and we sat there for a long time.

"I noticed an official over talking to Junior and I asked what he wanted. Junior said, 'They're going to send us to the rear, said we jumped the restart on the last restart.' Well, we had run 50 laps since the last restart. So why didn't they black-flag us then? Later I found out what the deal was. He had run over one of the Setzers (a popular racing family in the Hickory area), and after the red flag they decided they were going to put us in the rear.

"Well, there wasn't but 10 laps or less to run, so I told Junior to park the car. He pulled out and came around and stopped right in the middle of pit road. Those people in the stands went berserk. They were throwing bottles, rocks, and sticks, because they hated Earnhardts in Hickory, too. We were throwing stuff at the grandstand, and they were throwing stuff at us. Somebody had torn a fender off, so one of the guys on the crew grabbed the fender and ran out there and threw it toward the grandstands.

"There was some guy in the stands who stood up, and I could hear him holler just as good, and he said: 'Yeah, you redneck bastards from Mooresville, go on back home.'"

The incident is commemorated on the wall of Hargett's garage, where a piece of sheet metal from one of the cars he and Earnhardt Jr. campaigned now hangs high on the wall. The sheet metal has an inscription from Earnhardt Jr.: "The redneck bastards from Mooresville, Dale Earnhardt Jr.'s gang."

Myrtle Beach Speedway lies 150 miles south of Gary Hargett's garage in Marshville. Although there are dozens of short tracks in the Charlotte area, many within an hour's drive of Marshville, Dale Earnhardt chose Myrtle Beach Speedway as the place to officially begin Junior's career. It was a stroke of genius.

Gary Hargett raced with all three Earnhardts—Ralph, Dale, and Dale Jr. *Larry Cothren*

At the tracks closer to Charlotte—those in Hickory or Concord, for example—Junior was sure to be under an intense microscope, both from the public and from the media. After all, this was the son of stock car racing's biggest star and his every move would be scrutinized. Junior's older sister Kelley and half-brother Kerry were getting started in late-models at the same time, but at different tracks. Dale thought it was important for the three of them to race separately, each making his or her own way. At Myrtle Beach, under the tutelage of Hargett, Junior could focus on racing against some of the best short-track racers to be found anywhere.

"To me, that's the perfect race track for somebody to start out at," says Hargett. "They call it a half-mile, but it's a big race track, and it's got long corners. They're not real tight corners, and the race track is wide. If you get into trouble, you've got room to get yourself out of it, and you get used to running real fast down the straightaways, so it's an easy race track to learn on. Now, if you hit the wall, you tear up stuff because you're running so fast. But 10 cars can spin out on the straightaway and you've still got room where you can get around them without getting in a lot of trouble. You get used to the car getting out of shape a little bit, but you've got enough room to get it gathered back up too."

Yet, the potential that Hargett saw during Earnhardt Jr.'s street stock days at Concord needed some refinement. The natural ability—the feel for a race car, the willingness to mash the throttle, the on-track timing, all of it—was there. Raw but unrefined talent had been his daddy's trademark back in the 1970s, and it was the same with Earnhardt Jr. running Myrtle Beach and stepping up from street stocks to late-models.

"He ran good right off," says Hargett. "Like I said, he was the wildest thing you've ever seen on the race track. He would run sideways all the way around that track. I could not get it out of him, but I remembered Dale was the same way. He would run over people. If they were in the way, he would run over them. I remember Neil Bonnett was talking about that one time and said about Dale, 'That's the roughest boy I've ever seen in my life.' That was when he first started, and Junior was about the same way, though, for a while.

"But he was smart, although he would argue with you about what the car was doing. I mean, from day one, he would tell you exactly what he wanted that car to do. We didn't agree on it a whole lot, but I would try it just to show him that he was wrong. But he wasn't always wrong. He had a feel for a race car, and if I told him he could run wide open all the way around that race track, he would try it. And if I told him he had to get on the brakes at this point or this point, he would do that, too. He would do exactly what I told him to do.

"Still, when the race started, he was going to run pretty much the way he wanted to run. One thing that really impressed me right off the bat, and I didn't tell him and I know his daddy never told him, is that if that race car quit working somewhere on that race track—say he's chopping the bottom off and was real fast but it quit working down there—then he would find somewhere else on that race track that he could go and make that car run. He did that from the first year he ever raced, and he still does it now. If he told you he thought it was too stiff in the front or too soft in the front, most of the time he was right. He had a natural feel for what the race car was doing."

"He was a good race car driver," says Ernie Hammonds. "The first year Gary had to coach him through everything. He had the natural talent, but he was just so eager to go to the front that if you were in his way, you were going to move. Little by little, he got better at it. About halfway through the second year, he was a whole lot like, well, he reminded me of Rusty Wallace. He drove a lot more like Rusty than Dale Earnhardt."

The biggest obstacle for a young racer to overcome at Myrtle Beach Speedway is the intense competition. The Powell brothers—Robert, Charles, and half-brother Sean Graham—have ruled the roost at Myrtle Beach Speedway for a decade or so, and few have challenged their supremacy. Junior won just once in his three seasons there. The often-told story is that when Dale Sr. heard about the win, his first question was, "Did you beat the Powell brothers? Because if they weren't there, you didn't win anything."

For the record, the Powells weren't there on August 20, 1994, when Junior won his first late-model feature at Myrtle Beach Speedway. Ditto for his next win there, in 1996. "They gave Junior a hard time because he didn't win, but nobody wins on a regular basis and beats the Powell family at Myrtle Beach," says Hammonds. "Between the three, they've probably won 250 races at Myrtle Beach Speedway, or pretty close. Don't get me wrong, there are other people who've won here, but if you come here and you beat those three people, you've earned your money—no matter what your name is. They gave Junior a hard time because he couldn't, but it's not easy for anybody."

Robert Powell was nearing 30 when Earnhardt Jr. began racing at the beach. By the time Earnhardt Jr. came along, Powell had long been dominant on some of South Carolina's top short tracks, including Summerville, Columbia, Anderson, and Myrtle Beach. Powell was NASCAR Weekly Racing Series National Champion in 1988, the year he won 23 of 31 races at Summerville, and he later began winning regularly at Myrtle Beach.

"Junior gave Robert a whole lot of credit for his learning to drive a race car because he finished behind him probably seven or eight races a year—right behind him," recalls Hammonds. "We could get close enough to get second—or fourth or fifth behind the three of them—but we couldn't get by them. They were the three who beat him on a regular basis. It was still a lot of fun."

Powell says they had even more fun during Friday- night shows at another area track.

"I was just the general flunky . . . but I had a whole lot of fun, and I'll never forget it," says Ernie Hammonds of his days hanging out with Earnhardt Jr. at Myrtle Beach. *Larry Cothren*

"We had our best times racing at a little short track in Florence," says Powell. "It's a flat, worn-out, drivers' race track. He was good over there. You know, he was wild as heck anyway. We would run into each other and, I mean, we had fun. He told me one year, 'You know why I always run into you on the race track? You remind me so much of my daddy it makes me sick.'

"I knew he was going to be good, though. He wasn't scared of nothing. Wide open. He always handled himself good, too. He's probably one of the few racers I never really had a problem with. He's like me. You come here to race and you race. Rubbing is racing. That's the way I race, and people who can't handle that get mad. The two of us, we would rub and beat and bang and then get out and laugh about it. Because we knew we wouldn't wreck each other, just race.

"I remember one time we got rough, and I kind of got mad at him. I said, 'Man, are you crazy?' He said, 'What are you talking about? We were just having fun.' He ran into the side of me, knocked me to the top of the race track, and passed me. I would go back by him and he would get

The Earnhardts—Kelley, 16-year-old Dale Jr., Teresa, and Dale Sr.—at the 1990 Winston Cup awards banquet. *Stock Car Racing Archives*

into the side of me, knock me to the top of the race track, and go back by me. Finally, I just ran into him and knocked him about off the race track."

Despite the skirmishes, or because of them, the two developed a mutual respect. Earnhardt Jr. sometimes sought advice from his older competitor. "He would come to me a good bit," says Powell. "But Dale Jr. pretty much stayed to himself. We raced here together for three years. The first year I bet we didn't speak four or five words. I know how it is, or I felt I knew how it was with him. Everybody is always trying to ride him and all. I just took him as Dale Jr. That's the way I am. I think he took me the same way. He knew I didn't look at him as a rich kid or whatever, because I knew he struggled and he had to earn what he got.

"I never saw him where it was like he had to do this. You know what I'm saying? He would just take it as it comes. It's almost like, looking back now, he knew where he was going and he knew what he had to do to get where he was going. No matter how he did it, he was going to be there. Like I tell everybody, you've still got to be able to do it. I don't care how much money you've got, you've still got to be able to drive. And he could drive a race car."

"You were* not *going to beat him."

The speedway is a small part of the fabric of life in Myrtle Beach, which is one of the most popular vacation spots on the East Coast. The area is rich with tourist attractions—or distractions if you're young, single, and away from home for one of the first times in your life. Earnhardt Jr. and company came to race, but they were also intent on enjoying themselves.

Through the week, Earnhardt Jr. often stayed with the Hargetts, stretching out in a sleeping bag at their home near the race shop. He and Gary would prepare the car during the day, with Earnhardt Jr. showing an aptitude for chassis setups and decisions on changing shocks and springs on the car. Then on Saturday mornings, Hargett and Earnhardt Jr. would leave Hargett's garage and make the two-and-a-half-hour drive to Myrtle Beach. They would unload the car, turn laps during an early practice session, and hit the area attractions. That meant swimming on the beach or at a motel pool, playing video games, or riding go-carts. The group normally included Earnhardt Jr., Hargett, Ernie Hammonds, and one or two others who were helping with the car at that particular time.

Earnhardt Jr. and Hargett would occasionally spend time at a souvenir shop owned by Nick Lucas, part owner of Myrtle Beach Speedway. Earnhardt Jr. would sign autographs and hang out. Hammonds ran the souvenir shop for Lucas and it was there where he got to know Earnhardt Jr. and Hargett. "Junior would look through there and see if I had any new Darrell Waltrip stuff, because believe it or not Darrell Waltrip and Jimmy Means were his heroes, not daddy," recalls Hammonds. "That's pretty amazing. We would leave from there and go down on the beach and play those pole position games for hours and hours and hours, then ride up and down the beach.

"He would come back to the race track at about one o'clock and start doing his part of the deal. Sometimes he would wait until four or five, but by five o'clock he'd be there. We would stay there for the racing deal. Then when it was over, we would go to the beach and play the games, and go eat, then go on the beach and ride. He would spend a couple hours sometimes just riding up and down the beach, because he never had any freedom. I mean, this was the first freedom Dale Jr. had away from his daddy."

The autograph sessions at the Myrtle Beach souvenir shop were small, informal affairs compared to the ones Earnhardt Jr. encountered later on. "He was real shy, real, real shy," says Hammonds. "I went with him and did the first really big autograph session he did at the opening of a Wal-Mart in Charlotte—I think it was in 1993—where seven– or eight–hundred people showed up and fifty people came through there before he ever spoke. He would sign, but just hold his head down."

The shyness did little to mask a competitive streak and a level of determination that left an impression on both Hargett and Hammonds. It became obvious during the Myrtle Beach days that this was another Earnhardt who was going places in stock car racing.

"I didn't even put the Earnhardt into the deal," says Hammonds. "I knew as competitive as he was and as bad as he wanted to race that he would make it. Every time I went anywhere with him, we would get in a go-cart or play a video game with a race car on it—that's how I knew he was going to make it. That's all he wanted to do. And you were not going to beat him. You were *not*

It was clear early on that Earnhardt Jr. would have a lot to live up to. *Jerry Haislip*

going to beat him. We would go to the go-cart track and he was going to beat you, no matter who you were or when it was. And you couldn't beat him on the video games, and still can't."

The go-cart races often turned into slam-bang affairs, with Earnhardt Jr. giving no quarter.

"As a matter of fact, Junior put me in the hospital down there one Saturday afternoon," says Hargett, laughing at the memory. "He turned me over in the go-cart and broke my ribs. They had to carry me to the hospital and get me taped up. What happened was we were beating and banging and he spun me out, then drove up over the wheel and into my side with his front bumper. Then he got out laughing about it."

Occasionally, the nightlife around Myrtle Beach left its mark on Junior.

"I guess my favorite story is from when we had a big late-model race here one November," says Hammonds. "We had been out and had quite a few drinks the night before. Junior stopped on the track and said, 'I've got to have something to drink,' while the race was under caution. Gary told him, 'We can't bring you anything to drink out there.' He said, 'Get Ernie to bring me some water.' So I take him a cup of water out there.

"Kelley, his sister, is out there and she's hollering at me, 'Don't take him any water. He needs to learn better.' So I take him that cup anyway. Then he hollers back on the radio, 'Bring me some more.' So I take him some more. So then he hollers, 'I've got to have another drink.'

"Benny Parsons and his son Kevin were at the race. Kevin tried racing here for a little while and was here for that race. Junior said, 'Go by Benny's trailer and get me a Coke.' So I went over and said, 'Benny, Junior wants to know if you've got a Coke he can have?' He said, 'How are you going to get it out there?' I said I would take it to him. He said, 'I wonder what his daddy would think if I told him this story?' I said 'Please don't do that, Benny.'

"Luckily I was good enough friends with people at the race track where I could walk out and take him a drink—or I don't know what might have happened. They asked what I was doing out there and I said my buddy is going to die if I don't get him something to drink."

For the most part, the people who had close contact with Earnhardt Jr. during the Myrtle Beach days have little contact with him today, although Hammonds still talks to him on the telephone and sees him occasionally. The weekly phone calls Earnhardt Jr. used to make to Hargett when Earnhardt Jr. first started running the Busch Series have stopped, and the two haven't talked in the two years since Earnhardt Sr.'s death. "A month after Dale Sr. got killed, Junior was like a different person altogether," says Hargett.

Robert Powell, the short-track ace, saw Earnhardt Jr. in Darlington soon after Earnhardt Jr. moved to the Busch Series. Earnhardt Jr. called out to him as Powell walked by and the two spent a few minutes together. They've talked very little since then.

Hammonds relishes the time he spent with Hargett and Earnhardt Jr. and is modest about his role in Earnhardt Jr.'s Myrtle Beach experience. "I was just the general flunky," says Hammonds. "I was the general flunky who tried to get Junior where he needed to be at the right time, but it was fun. It was really neat. I mean, I can say I hauled Earnhardt's kid around, and up and down the road. I wouldn't trade it for anything I've ever done in my life. I didn't make any money, but I had a whole lot of fun and I'll never forget it."

2

MOVING UP

BY LARRY COTHREN

From Myrtle Beach to the Busch Series

All eyes were on Dale Earnhardt Jr. Everyone wanted to know if he would follow in his father's championship footsteps. *Paul Melhado*

By 1996, Dale Earnhardt Jr. and team owner Gary Hargett had parted ways. The two won just one race together, at Myrtle Beach in 1994, before going winless in 1995. Hargett's Marshville shop served as the center of his and Earnhardt Jr.'s racing activity for three seasons, but Earnhardt Sr. wanted the team to be headquartered in Mooresville for the 1996 season. Hargett turned down Earnhardt Sr.'s offer of a job in the Mooresville shop. After years of being his own boss, Hargett felt he couldn't take orders from someone else, particularly Earnhardt Sr.—someone Hargett knew to be every bit as headstrong as Hargett himself.

So, during the 1996 and 1997 seasons, Earnhardt Jr. drove a car built in his father's shop in Mooresville. The shop was small in comparison to the mega-shops prevalent today, but it was the forerunner to the complex that now houses Dale Earnhardt, Inc. Earnhardt Sr., just beginning to really make his mark as a team owner, was fielding a competitive Craftsman Truck Series team and a fledgling Busch Series team out of the Mooresville shop. Ron Hornaday had won six times and finished third in points while driving Earnhardt's truck during the inaugural truck series season in 1995. Hornaday then won four races and claimed the series title the next season. By 1997, with Steve Park as a rookie driver, Earnhardt's Busch Series team was ready for another full season. Earnhardt Jr.'s late-model cars were built in 1996 and 1997 by some of the same crewmen who built the trucks and the Busch cars, with Earnhardt Jr. providing assistance.

Earnhardt Jr. competed at several different tracks—Myrtle Beach, Nashville, and I-95 Speedway in Florence (South Carolina), among were not significantly different from what he experienced in Hargett's car; the effort was still just a step away from consistently visiting victory lane.

"We ran 16 races at I-95 [Florence] in 1996," recalls Earnhardt Jr. "I finished second 14 times. We won one race and the engine blew up in the other race. We were mainly just bouncing around, from Nashville to Florence to Myrtle Beach."

He added a win at Myrtle Beach to the one at Florence in 1996 before going winless again the next season. The Busch Series was beckoning Earnhardt Jr. by then, however, as he started his first Busch race at Myrtle Beach Speedway on June 22, 1996. He then ran eight Busch races in 1997. His record in those events was nothing spectacular, with only one top ten, a seventh place at Michigan on August 16, 1997. There was certainly no hint of the Busch success that would follow over the next two years.

By the end of 1997, his late-model career was over, although Earnhardt Jr. can still recite his record in NASCAR's Weekly Racing Series: 159 starts and 3 wins.

"It was a good time," he says. "I learned an awful lot about driving and quite a bit about race cars. That's really the only time in my career where I worked on my own equipment, and for the most part did probably 80 percent of the job on my car, getting the car down to the track and driving it. Doing those things was fun. It was great because you were on your own schedule pretty much and you knew what you had to do to get the car ready for the next race.

"You don't really find Winston Cup or Busch Series drivers these days coming from the late-model ranks as often as you used to. You used to get a lot of late-model drivers out of the Winston Racing Series into the Busch Series, but you don't anymore. I think it's overlooked now because it's not a highly publicized series like it used to be. I wouldn't trade the four years I raced the Winston Racing Series for anything because I think it was a great, great foundation for the driver I became in the Busch Series."

The driver he became in the Busch Series had a rude introduction to the tour in the season-opening race at Daytona in 1998. His car suffered a broken drive shaft on a pit stop (driver error) and a wild flip on the backstretch (wrong place at the wrong time) on the way to a finish of 37th. Yet he recovered quickly in the next five races—grabbing 16th place at Rockingham, 2nd at Las Vegas, 3rd at Nashville, 10th at Darlington, and 2nd place at Bristol—before getting his first Busch Series win at Texas on April 4, 1998. He then had a total of seven wins that year. Earnhardt Jr. finished his rookie season by winning the Busch Series championship with the team that Steve Park put in victory lane three times the previous season.

Earnhardt Jr. had clearly arrived as a NASCAR star by the time *Circle Track*'s Bob Myers interviewed him during the middle of Earnhardt Jr.'s 1998 rookie season.

DALE EARNHARDT JR.

The Younger Earnhardt Is Making a Name of His Own

BY BOB MYERS
From *Circle Track* September, 1998

This article is an interview of Dale Jr. conducted by Bob Myers.

As Dale Earnhardt's 23-year-old son and driver of the number 3 ACDelco Chevrolet he owns in the NASCAR Busch series, do you feel like you're living in a glass showcase with everybody watching?
Sometimes I do, but I have to concentrate on what's important, and that's the race team.

Obviously you are proud to be an Earnhardt, but is there pressure on you to live up to your father's tradition of excellence?
Not really…I want to be my own person and make a name for myself, but there's no pressure to be as successful as my father.

Do you sometimes wish at the race track that your name wasn't Earnhardt?
No. Some people think there are disadvantages and pressure to being an Earnhardt, but I think there are more advantages than disadvantages.

Do the assets—money, experience, advice—available to you put an extra burden on you to perform?
Naw. It makes it easier. It presents the opportunity to run good. The only thing new about the team that was successful with Steve Park last year is me. We wanted to try to pick up where Steve left off [3 wins, 12 top fives, and 3rd in points].

Moving to the Busch series full-time this year, how did the ride in the number 3 Chevrolet come about?
For a couple years, I had thought about the possibility of driving that car. I don't think dad thought I was ready. Tony Eury [veteran crew chief] and the guys on the team wanted and expected me to drive. There were lengthy discussions and meetings, and I don't think daddy made the decision

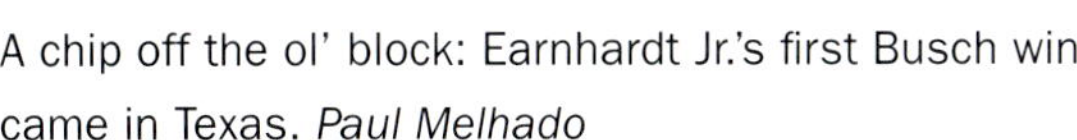

A chip off the ol' block: Earnhardt Jr.'s first Busch win came in Texas. *Paul Melhado*

Daddy Dale celebrates at Daytona. *Sam Sharpe*

until a month before the [opening] race at Daytona. And he probably wondered about me during the first three or four races.

What did he tell you at decision time?

He still hasn't told me anything face-to-face. We have no agreement or contract.

After winning your first Busch Grand National [BGN] race, a major one at Texas Motor Speedway in April, in only your 16th BGN start, and after logging four top fives, including two 2nds in eight starts before a wreck and 32nd place at Talladega cost you the points lead, have you exceeded expectations?

So far, on the general timeline, we've accomplished more earlier than I thought we would. I wanted us to be real consistent, but I didn't think that would come three races into the season. I thought I would need more time to build confidence.

Did you expect to win so quickly?

No. But I knew we were capable. We could have won earlier. [He finished second at Las Vegas and Bristol.]

"When he has all his little ducks in a row, he doesn't want anybody messing with them. I'm kind of like that," says Dale Earnhardt Jr. *Paul Melhado*

> **"He has done so much for me in racing and life. I feel compelled to drive for him and try to win as many races and championships as I can, to try to pay him back."**
> *—Dale Earnhardt Jr. on his father*

In fact, what are your expectations this year?
To finish in the top five consistently. Do that, and we could win the championship without another win.

Was the warm embrace by your proud dad—reminiscent of golf phenomenon Tiger Woods and his father at the 1997 Masters—after the Texas victory an emotional moment for you?
The most emotional of my life. It stirred memories of the years I had tried so hard to earn my dad's approval. Maybe that did it. It really was a proud moment for him, to show that much excitement and happiness over something that I had accomplished.

Does your father coach you on the radio during races?
Somewhat. He sees opportunities and situations start to develop way before I do. He's been there, done that. He recommended a four-tire final stop in the Texas race, and we decided that was the route to go. That call was a major factor in the win. During the final six laps, the radio was quiet.

You've said your dad's coaching makes you nervous. Why?
Because I know how much he wants me and the team to do well, and because he's my father. I'm constantly concerned about being ridiculed or criticized for something, because I know he is the best, at least in my eyes. It means a lot for me to do well, and hearing him on the radio brings all that to reality. I can get on the race track and forget everything else, but when I hear that voice, I know he's watching my every move.

What's the most important thing he has instilled in you?
Determination and drive.

What did your daddy say to you after you damaged the drive shaft in the pits and then crashed spectacularly during the Daytona 300, the first race of the season?
Nothing. He called on the radio and told us to calm down, just fix the car [drive shaft] and for me to get back in the race for drafting experience. After I crashed, I was sure he was going to chew my head off.

How did you feel after he won the Daytona 500, in his 20th try, the next day?
That was unique. I experienced the same emotions he did, because I knew of all the near misses he had over the years. I will always regret I wasn't there to enjoy it with him, because it's something he really wanted. I watched on television. I had come home because I didn't feel too well after my wreck.

Describe your driving style.
Aggressive, more so at times than it should be, relative to the way NASCAR runs the series. NASCAR is less tolerant of aggressive driving, and you can't be as aggressive as my dad used to be. I enjoy that kind of racing. I hate watching a race that looks like a bunch of toy soldiers marching around. Fans like action, even if their favorite driver gets bumped around or spun out.

Earnhardt Jr. finished 16th in the ACDelco 200 at Rockingham in October of 2000; he carried the ACDelco colors full-time the next season. *Nigel Kinrade*

How much did that first victory mean to you and your team, led by Eury and sponsored primarily by ACDelco?
It meant a lot. It was neat to finally show the ACDelco people that they had made the right decision, to trust a rookie as young as I am. Regarding the team, I grew up with Tony about as much as with my dad. Tony might not realize it, but he taught me a lot and has been a big influence on me. It was cool to win the race as much for him as my dad.

Fresh from the late-model ranks, Earnhardt Jr. hit the Busch Series in 1998. *Harold Hinson*

Were you aware of fans at Texas cheering practically every car you passed and going crazy when you got under Nemechek? If so, how did that make you feel?

Not during the race, but once I got out of the car in victory lane I sure was aware. I'd gotten some idea of how many fans were supporting me from the cheering when drivers rode around the track during introduction. I think we might have picked up a few more fans that day. It was a thrill.

You're always in the spotlight because of your name, but has the media attention increased since your first victory?

Yeah. I'm being pulled right and left.

Do you enjoy, and how important is, media awareness—doing interviews, mixing with fans, and keeping other commitments off the track?

I enjoy autograph sessions the most because I meet the fans, and they are a lot of fun. Even though I am Dale Earnhardt's son, I've always been a big fan. I feel I can relate to fans because they're my kind of people. Some of the other stuff—dinners, parties, and meetings—I'm not too cool about. I'm not much on dressing up, especially in a tuxedo.

While he waits to turn laps, Earnhardt Jr.'s crew works on the car during his Busch Series debut at Daytona in 1998. *Harold Hinson*

Earnhardt Jr. was strapped in for a Busch Series race at Bristol Motor Speedway in August of 1997. *Harold Hinson*

In 1998, Texas Motor Speedway was the scene of Earnhardt Jr.'s first win in the Busch Series. *Harold Hinson*

After nearly two decades in the spotlight, your father seems to be somewhat shy. Are you that way?

I don't know if it's shy, maybe we just ain't good at putting words together. If we could express ourselves better, without offending anybody, we'd talk and try to get our point across more often. It's so easy to hurt people's feelings.

At 6 feet, 165 pounds, you have your father's build, some of his features, and apparently his racing genes. What other similar attributes do you possess?

I don't know. He's not one who likes to be imitated or copied. If I were to say I am like him in any way in this article, he wouldn't be too happy with me. He's very possessive and territorial. When he has all his little ducks in a row, he doesn't want anybody messing with them. I'm kind of like that.

Earnhardt Jr. competed in eight Busch Series races in 1997. *Harold Hinson*

You don't have his menacing mustache. Do you plan to grow one?
Naw. I don't even know if I can. I might grow a beard during hunting season.

After racing in the Late Model Division of the Winston Racing Series from 1993 to 1996, winning three starts, with 59 top-five and 90 top-ten finishes—and after competing in eight Busch races last year, with a best finish of seventh—were you convinced you were ready for the Busch Series full-time?
Yes. In the Winston Racing Series, I helped put together, worked on, and set up my cars. I learned from my mistakes, and that's paying off now. I wasn't a dominating driver, didn't win many races, but I was consistent.

What's your opinion of Winston Cup drivers competing in Busch races?
I like it. Having a chance to race against them is like taking a class in school. If you're in the top five at the end of a race, you've made an A+ in the class.

Even though your father provided resources and guidance, he didn't serve everything on a silver platter. Haven't you had to work for what you have?
Yeah. He supplied the cars and equipment, but there was sponsorship for the program and everything wasn't out of his pocket. He said, "Here's an opportunity. Here are some cars. Race them. In three years, we'll see how you're doing." At least I learned enough to move to Busch.

A Busch Series ride at Lowe's Motor Speedway in 1997 bore the number 31 and a resting Earnhardt Jr.
Harold Hinson

Is there a timetable for advancing to Winston Cup?
Not necessarily. What I'll probably do is run two or three more seasons in Busch, at least through 2000, then take the number 3 ACDelco team to Winston Cup. I think the team would be competitive there now.

Would you like to continue driving for your father, particularly after he retires as a driver?
I think so. I believe that's what he'd like me to do. He has done so much for me in racing and life. I feel compelled to drive for him and try to win as many races and championships as I can, to try to pay him back.

How many more years do you think he will drive?
Probably a couple more years, or maybe a year or two after I join Winston Cup.

Do you think about accomplishing what he has: 7 championships, 71 victories, and a motorsports-record $31.8 million in race earnings as of this writing?
Yes. And I think it's possible. The number of races and purses is increasing as Winston Cup grows.

You grossed a Series-leading $249,405 in prize money in the first nine BGN races this year. How much does money mean to you?
Not a whole lot, because I don't see much of it. I draw a salary and live on it. Like other drivers, I get a percentage of the race earnings, but that goes into a special account. I don't see that money. I like it that way. I know I have money in the bank to pay my taxes and to invest. Having more money hasn't changed me as a person. That can happen when a pile of money is dumped in your lap.

At what point in early life did you know that you wanted to make racing a career?
Driving race cars was a childhood fantasy. When I was 15, Concord Motor Speedway started a Street Stock Division. A newspaper that was lying on a table in the shop had an ad or story about the new division. I walked in after school and [brother] Kerry was there. Dad pointed to the paper

A year before this 1998 photo was taken at Lowe's Motor Speedway, Earnhardt Jr. was still turning wrenches on his late-model ride. *Nigel Kinrade*

and said, "You boys ought to take a look at that and get interested." That's when I had my first chance. I had wondered for a long time if I could do it. I sold my go-cart for $500, bought a car at a junkyard for $200, and built a race car with the other $300. I was fourth in the first race I ran. Kerry drove the car until I got old enough.

At what age did you start racing go-carts?
When I was 13, and only for a year. My dad didn't like carts because they didn't have seatbelts and roll bars, and I came flying out of them quite often.

Did the fact that you are an Earnhardt influence your decision to race?
Yeah. I saw firsthand how competitive and enjoyable racing was and what it was all about.

Did you tell your father you wanted to race, or did he ask you first?
I had told him a couple of times I wanted to race. When I was 13, I told him I didn't think I would make it to 16, that I just couldn't stand waiting and worrying. He wanted Kerry and me to get involved, but he didn't push us.

Are Kerry and your sister, Kelley, continuing to race?
No.

Tires smoked during a pit stop at Daytona. A broken drive shaft and a wild, tumbling crash on the backstretch spoiled Earnhardt Jr.'s debut at the track. *Nigel Kinrade*

The crew made hasty repairs at Talladega after the rookie driver crashed. *Nigel Kinrade*

Steve Park (right) was the driver for the ACDelco Chevrolet before Earnhardt Jr. (talking to Dick Trickle) took over. *Mike Gurnsey*

Did you hang around your dad's shop and go to the races when you were a kid?
Yeah. From 1981 until I started driving myself.

What other interests did you have while growing up in Kannapolis, North Carolina?
Practically none.

Where did you attend high school, and did you participate in sports?
I played a little soccer during my second- and ninth-grade years. I went to several schools: Christian, private, and military…because I was a bad little kid. I wanted to drive race cars, not fool around with anything else. A year and a half in a military school did more good for me than anything else. That was a wake-up call. I finally graduated from Mooresville Senior High.

How important is at least a high school education to a race driver?
Very. School is much more than book learning. It teaches you to deal with people, and if you can't do that, you're not going very far in racing.

Where do you reside?
I have a double-wide [mobile home] across the street from the shops. I do the cleaning and laundry. I enjoy working in the yard and have planted some bushes.

Earnhardt Jr. and Dale Jarrett chatted prior to a Busch Series race at Michigan in 1997. Earnhardt Jr. finished seventh—for his best run in seven starts that season—while Steve Park won. *Nigel Kinrade*

What type of person are you?
Real laid-back. I had to develop that style. In the 10th grade, I started getting bad stomach pains. This went on for a time. I worried myself sick about doing well in school and pleasing my father. Two years ago, I quit worrying about a lot of things, particularly what other people thought.

You're single, so does that mean you're an eligible bachelor or do you have a steady girlfriend?
Very eligible.

What leisure activities do you enjoy?
I enjoy hunting once or twice a year, although I'm not into that as much as my father is. I fool around with remote-control cars and play on my computer [for] hours on end.

What sort of lifestyle do you lead?
Very simple, nothing special or fancy. For example, I enjoy riding around in my pickup truck and watching TV. I like jeans, T-shirts, and sneakers, old country and rock-and-roll, solid gold soul, and comedy movies.

"Little E" helps push his Busch car to the starting line.
Warren Melhado

Where does your mother, Brenda, reside, do you see her often, and what does she say about your racing?

She has remarried and lives in Norfolk, Virginia. I talk to her on the phone from time to time, and she visits us at least once a year. I love her to death. She's really cool. I lived with her after she and daddy were divorced in 1978, until I came to live with my father and stepmother, Teresa, in 1981. Teresa has been a mother figure to me as well, and I'm grateful to her for everything she has done. A lot of people don't know that Teresa isn't my real mother.

You never knew your late grandfather, Hall of Fame driver Ralph Earnhardt, did you?

No.

Earnhardt Jr. quickly became a fixture in the Busch Series. *Nigel Kinrade*

Is grandmother Martha Earnhardt proud of you?
Oh, yeah. I'm still her little grandson, and she's nervous about me getting hurt. But I think she has more confidence in me now.

Are you concerned about getting hurt?
Naw. Not at all.

Would you say you are very close to your family?
Yes. I've gotten closer in the past couple of years, because I think I've shown them that I'm not the spoiled-rotten kid I was when I was 10–12 years old. That's part of growing up.

Dale Earnhardt Jr. is shown pausing prior to racing at Lowe's Motor Speedway. *Nigel Kinrade*

Are you generally pleased with your racing career to this point?

I think so. I have a killer instinct to win races, but if I don't win, I won't worry about it. If I finish good enough to make the car owner and sponsor happy, and when I get home have enough money to pay the electrical bill, that's fine with me.

What are your goals in racing?

To win a Busch and a Winston Cup championship. And it's important to win a Daytona 300 and a Daytona 500. Anything else will be a bonus.

Anything you wish to add?

I could go on and on about all the people who have helped me get where I am, but there's one I don't want to forget, Gary Hargett. He was my crew chief and mechanic my first three years in late-models, and I raced out of his shop in Marshville, North Carolina [about 35 miles east of Charlotte]. He worked with Harry Gant and my dad in the 1970s. He was the grandfather I never had. When I started driving for him, I was young and self-centered. But he and his volunteer crew gave me no breaks because of who I was. He taught me how to deal with people, how to get people to help you, and how to hold groups together. He used a lot of common sense. One of the hardest things I've done was to leave him. But it was necessary, when space became available, to move my race cars from his shop to my dad's operation in late 1995, so that I could work on them daily and learn more about them.

THE NEXT DALE EARNHARDT

3

From *Stock Car Racing* magazine, September 1998

He's Here, and He's Driving Daddy's Car

There are enough expectations to fill a transporter when you grow up as the namesake of a genuine motorsports icon. When *Stock Car Racing* caught up with Dale Earnhardt Jr. for this September 1998 cover story, he was 23 and had grown accustomed to the baggage that came with being an Earnhardt and driving a stock car for a living. He was also well on his way to carving quite a name for himself as a Busch Series competitor.

By the midpoint of the 1998 season, Earnhardt Jr.'s rookie year in the Busch Series, he had claimed his first two Busch wins, at Texas and Dover, and was in the midst of what would become a seven-win, title-earning campaign. It was quite evident that this kid—one year removed from the late-model ranks—carried more than just his father's name.

Earnhardt Sr. offers a little fatherly advice to his rookie driver. *Nigel Kinrade*

Dale Earnhardt Jr. was born with a surname which symbolizes action and decisiveness. He has done all he can to live up to it, yet he responds to the pressure with surprising patience and maturity.

"It's real early in my career," says Dale Earnhardt Jr. "I think a lot of people need to realize that."

Dale Jr. is but 23 years old, and a few months into the professional phase of his race-driving life. At this young age, he has several memorable performances to his credit. One April Saturday in Texas, he charged from behind after a late pit stop and picked a victory—his first in NASCAR's Busch Series—out of Joe Nemechek's pocket.

And how about Dover at the end of May? That was the race where young Mr. Earnhardt came onto pit road a little too fast, spun his car out of the lead, and then seemed so determined to make up for his blunder that he blew the doors off the rest of the field and won again. Patience, hell.

All Dale Earnhardt Jr. has done in the 1998 season is grab one of the toughest leagues in the sport and wrestle it to the ground. In the season opener at Daytona, he qualified third. Seven races into the schedule, in just his 16th start in the Series, he won at Texas. Dover was this year's start number 13. He has made running up front look like it was his birthright. And maybe it was.

There were few—very few—embarrassing moments like this Charlotte spin where Earnhardt Jr. is heading toward the Porta-John. *Dale Grubba*

Dale Sr. remembers giving his son this advice: "You don't have to do what I've done. Just go out and do what you can do. Just win races, and you'll do fine." Clearly, Dale Jr. listened. To say that he is doing fine would be the motorsports understatement of 1998.

He has become, hands down, the most popular driver in the Busch division. The official award won't be handed out until season's end, but if you listen to the crowd reaction at any given race you can see that the thing is a lock. And at every stop on the Series, he is the favorite subject of the local media. In fact, the drill for interviewing Dale Earnhardt Jr. at the racetrack has become much the same as it is for interviewing Dale Earnhardt Sr.: make an appointment, show up on time outside his transporter, do your interview, and make room for the next guy. It is the only way everybody will get their shot at him. The Busch Series, which has seen its share of young drivers so eager for ink

Earnhardt Jr.'s crew services the car on a pit stop at Atlanta. *Nigel Kinrade*

Earnhardt Jr. leads the pack down the frontstretch in Michigan. The popularity of the third-generation Earnhardt helped fill the grandstands for Busch Series events.
Nigel Kinrade

that they would wax a writer's car in exchange for a story, has never had a newcomer as hot as this one.

Maybe NASCAR hasn't, either. When Jeff Gordon arrived in the Busch Series in 1991, only those who had seen him in open-wheel cars had any real inkling that he was anything special; initially, a lot of stock car folks wrote off Gordon as a wild child who might never stop crashing. Before Jeff, the last wunderkind who generated any real media splash even before he reached the Winston Cup level was Kyle Petty, who was dubbed a "prince"—well, his father was a King—by the full-time beat writers after his 1979 ARCA [Automobile Race Car Association] victory at Daytona. But face it, you could seat all of those full-time beat writers in a pretty small pressroom in 1979.

Dale Earnhardt Jr. comes into the spotlight at a time when NASCAR has never been bigger, and he comes bearing its biggest name. Life as a "son of" is never easy and being Billy Earnhardt or Bobby Earnhardt would have been tough enough. Being Dale Jr., however, has to represent a whole

Dad looked on proudly in 1998 while Earnhardt Jr. was being interviewed after an exhibition race against Jeremy Mayfield at Lowe's Motor Speedway. *Harold Hinson*

different level of tough. Fortunately, it was hung on a pretty tough specimen; when I inquire about the burden of his name, he says, "It's never really been a problem."

"You don't ever feel like it's a lot to live up to?" I ask.

"No. I mean, I know I'm only going to be as good as I can be. All I can do is go out there and do my best."

He shoots a grin, which is not unlike his daddy's, slightly crooked and full of confidence, and reminds me that after 23 years, he has sort of gotten used to being Dale Earnhardt Jr.

"I don't know any different," he says.

The grin is not all that these Earnhardts have in common. On the youngster's best days, when he races with the blend of brains and brawn that he displayed at Texas and Dover, their driving styles look almost interchangeable.

"He's a competitive little driver," says the father of the son. "He likes to get aggressive."

The only problem, according to Dale Sr., is that there are days when the boy's maturity is still a lap down to his desire. He points to the springtime events at Nazareth and Charlotte, and says, "He had the best race car two races in a row, and he wrecked it two races in a row. You've got to be able to be aggressive and use your head, too."

That is enough to raise the eyebrows of Tony Eury, crew chief for the Busch Series arm of Dale Earnhardt, Inc. Eury has chummed around with Dale Sr. forever, "I was probably 11 or 12 when I met him," and they were working together on Earnhardt's short-track cars as long ago as 1976. He has

Earnhardt Jr. and Matt Kenseth became friends while competing in the Busch Series. They're shown before the next-to-last race of the season at Atlanta in November 1998. Earnhardt Jr. edged Kenseth for the Busch Series title by 48 points one week later. *Nigel Kinrade*

known Dale Jr. quite literally since his birth in October of 1974. If anybody on the planet qualifies as an expert on drivers named Dale Earnhardt, it is Eury.

"A lot of people think Dale Jr. is a hard racer," Eury says. "But when his daddy first started, he was worse."

The elder Earnhardt, Eury claims, ". . . was his own worst enemy. He probably should have won a lot more races than he did [in the beginning], but he tore up his equipment before the end of the race."

Those long-ago mistakes, blurred as they were by Earnhardt's 1980s rise to big-league glory, were not forgotten by the man who made them. They became, Eury believes, part of Dale Jr.'s education: "His daddy has told him about all that stuff," the mechanic says.

The lessons took hold. Dale Sr.'s protests aside, young Earnhardt has gotten himself into no more trouble than several other fast Busch Series rookies have over the past few seasons.

"I'll get a little aggressive when the situation calls for it," Junior says. "But I try to be patient as much as I possibly can."

Dale Earnhardt Jr., at 23, is in many ways older and wiser than his dad was at the same age. If anything, he resembles the racer Dale Sr. became more than the racer Dale Sr. used to be. Sometimes the similarities are eerie.

"I haven't tried to raise him to be like me," the father says when I ask about this. "I've tried to raise him to be his own person."

In that, he has succeeded; it's just that the person Dale Jr. became has an awful lot in common with Dale Sr. Both of them stick close to their cars and their teams at the track, focusing on the job at hand and only occasionally socializing with competitors. Both of them have an uncanny ability to isolate themselves in a crowd; even as nosey onlookers besiege their garage stalls, they seem somehow alone at the center of it all. Both of them are essentially shy, and during interviews they often seem worried that they are giving too much of themselves away.

> **"You don't have to do what I've done. Just go out and do what you can do. Just win races, and you'll do fine."**
> *—Dale Earnhardt Sr. to Earnhardt Jr.*

Earnhardt Jr. insists that there are plenty of differences between them, and that one reason he has been doing so many interviews is "so people can see more of my personality. Because whatever they don't know about me, they assume it's pretty close to what Dad is like."

But even as his own persona emerges, there is no escaping the fact that it really is pretty close to his dad's. Exhibit A: I am pressing Dale Jr. about the ways in which he and his father are alike, and he says, "We both like to get away, like to be left alone. Have a good, peaceful day by yourself, you know? We're very, uh, independent."

That brings to mind something I heard Dale Sr. say years ago about his own father, the late NASCAR sportsman champ Ralph Earnhardt. "He was pretty independent, pretty quiet, did his own thing," Dale Sr. said that day. "He wasn't the type who depended on too many other people."

So maybe all of it, including the talent and the independence, is in the Earnhardt genes, passed down from Ralph to Dale to Dale Jr. In his most open moments, Dale Sr. will fondly recall sitting in the grandstands at various Carolina bullrings, watching Ralph Earnhardt work his magic in traffic, and then quizzing him about his winning moves. Dale Jr. has heard and read those stories enough to have developed his own questions about Ralph Earnhardt, who died the year before the youngest Earnhardt was born.

"I'm interested in what kind of person he was," Junior says. "You know, what kind of personality he had, whether he was outspoken or not, things like that. I like to find out about that as much as I can."

I ask Dale Jr., "Do you think you studied your father as closely as he studied his?"

He sighs. "I don't know for sure how closely he studied ol' Ralph. But I've paid a lot of attention to Dad. I've watched him make good moves and bad moves, and I've tried to learn from both."

Curious, I throw the same question to Dale Sr.: Does he suppose his son watched him the way he watched his own dad? I am surprised to catch Earnhardt looking rather curious himself.

"I don't know, really," he says.

Now there is a hint of melancholy on the champion's face. He says, "I went to the races and watched my dad week after week, and Dale Jr. has done that, too. But I hung around the shop more when I was a kid than Dale Jr. did."

True enough: Dale Jr. had a much different childhood than his father did. Dad grew up knee-deep in race car parts, tagging after Ralph. Junior was born when Dale Earnhardt was flat broke and finding wedded bliss as elusive as racing success. "I struck out on marriage a couple times," says Dale Sr. Little E was relegated by his parents' divorce to the fringes of the sport. Living with his mother, his first exposure to racing consisted of listening to events on the radio, "probably in 1978. Dad was getting just a couple starts in Winston Cup."

He says quietly, "It was kind of tough being away from him."

It wasn't until 1982, when Dale Sr. had become a NASCAR hero, married his third and present wife Teresa, and assumed custody of Dale Jr., that his son "got the chance to be around him, and find out more about what he was like."

The number 3 was a natural fit for the son of the man who made the number synonymous with hard driving and stock car success. *Nigel Kinrade*

It was no surprise that he chose to answer the family calling. Junior's first race came in a go-cart "when I was about 12." The next step was a "celebrity" race in Legends cars at Charlotte, running against the offspring of some of his dad's Winston Cup rivals. By 1994, at age 19, he was a regular in NASCAR's late-model stock car class. His home track was South Carolina's Myrtle Beach Speedway, but before long he branched out, hopscotching around the region "just to learn about different bankings, different [track] lengths, different speeds, different corner entries and exits."

I say to Junior, "Could you tell right away that you had the ingredients to be good?"

He gives me his daddy's crooked and confident grin again, "Yeah. I knew I could drive a good car well."

The job of keeping young Earnhardt in good cars initially fell to Gary Hargett, a veteran wrench who years earlier had labored on Dale Sr.'s Saturday-night specials. In Hargett, Junior found a mentor: "I was just a wild buck, but he taught me about race cars, taught me how to handle myself, taught me how to work with people."

This was, as you might guess, part of a plan. Dale Sr. says, "When I started, I drove for other guys, too. I had already learned a lot by working with Dad, but there's still that learning process where you need to find things out for yourself by working with different people like Dale Jr. did."

Although not as intimidating as his father's black number 3 Chevrolet, Earnhardt Jr.'s blue number 3 led the Busch Series in wins (7) and money won ($837,065) in 1998. *Nigel Kinrade*

It helped young Earnhardt the same way it helped Dale Sr.: the more he learned, the faster he got, and the more he learned.

He remembers nights making his first big chassis decisions himself, running great, and thinking, "Hey, I'm getting a hold of this." Soon, after learning by a method Dale Sr. refers to as "trial and error, learn-as-you-go," Junior was calling his own shots.

This was Earnhardt's way of dropping the boy out of the nest. He had taken the same tack with his oldest son Kerry (now 28) and daughter Kelley (25) when they began racing late-models a few years earlier; all three kids worked on their own cars. In fact, all three kids worked, period, because Dale Earnhardt did not believe in free rides. If his children wanted to go racing, that was fine, but they would have to earn their keep while they did so. They could hire on at one of the family businesses, or they could find jobs elsewhere, but they were damn sure going to work.

"I wanted them to be able to handle their own affairs," Dale Sr. says. "I wanted them to be able to handle life."

Richard Childress offers advice prior to Earnhardt Jr.'s debut at Daytona. *Harold Hinson*

Junior landed in the service department at Dale Earnhardt Chevrolet and supported his late-model Chevy by performing oil changes on the local citizenry's Cavaliers and Caprices.

Today, Dale Jr. has what he calls "a lot of respect" for Dad's earn-your-keep approach to raising children and race drivers.

"When it was going on, I didn't like it much. But it is the best way. It makes me take a lot of pride in my racing, and a lot of pride in what I've done."

He pauses. "And a lot of pride in why I'm sitting here."

He means, of course, sitting in the DEI fleet of Busch cars. He had been around those cars for years, had watched Jeff Green and then Steve Park drive them, and, as any racer would, had longed to drive them himself. But his father's no-free-ride policy stood: simply being an Earnhardt didn't entitle you to the keys of the family Busch cars.

"I didn't know if I was ever going to get the chance [to move up]," Junior says.

But the chance did come, in June of 1996 at Myrtle Beach.

"That was a place he had run a lot," Dale Sr. says, "so we felt it would be a good opportunity for him to get some experience in these Busch cars."

Tony Eury Sr., uncle to Earnhardt Jr., provided a steady hand during their first season together in the Busch Series and later moved up to Winston Cup with Earnhardt Jr. *Nigel Kinrade*

Green was the team's lead driver at the time, and Dale Jr.'s ride was little more than a decent backup. But he qualified 7th and, after spinning early on and losing a lap, soldiered home 14th, a result which didn't come close to reflecting the show he put on.

"We passed cars all race long," Junior says. "We could have maybe finished in the lead lap, but I wasn't running for points, so I never got up to the front of the [restart] line under caution. I stayed in the back, out of everybody's way. Then I'd pass 'em again. It was like we passed everybody there, over and over."

His father watched it on television from Michigan, where he was running in a Winston Cup race the next day. Seeing his son blow past some of the toughest racers in the Busch Series, he says with a gleam in his eye, "was a good feeling."

Good enough, apparently, to book the lad for eight Busch events in 1997, beside the team's rookie star Steve Park. Junior's starts came at tracks ranging from the Nashville half-miler to the Watkins Glen road course and the 2-mile California Speedway. It was, Dale Earnhardt admits, an audition: "We were trying to give him a feel for [the series], and give ourselves a feel for him." On the strength of a seventh-place finish at Michigan, an outside-pole qualifying run at Bristol, and a top-ten start at the Glen, Junior passed.

Tony Eury says, "He never had the kind of equipment we had for Steve; Dale Jr. always had the second-best car, always had an engine that wasn't as good as ours, always worked on it himself, just him and a couple of young guys from his late-model team. But he managed to be pretty competitive."

Along the way, Park won three races and became the obvious choice to drive the new Winston Cup entry DEI would field in 1998. That left an open seat in the Busch car. Dale Sr. says, "We were like, 'OK, do we put in someone with experience, or do we take a chance with somebody new?' We had taken chances with Jeff Green and Steve Park, and they both did a good job. So we said, 'We'll just take a chance on Dale Jr., too.'"

Junior jumped at the offer. "I knew the team was capable," he says, "because they won with Steve. I knew all the ingredients were there to run well. We just had to find out whether I could do it."

I ask him if he wasn't just a little bit nervous about the whole thing, and he shoots back a look that tells me exactly how dumb a question this is.

"Of course I was nervous. But I wasn't going to turn it down."

The move put him under the watchful eye of Eury, who is schooling Dale Jr. just as Gary Hargett did in the late-model days. It also teamed him with Eury's son Tony Jr., who, in addition to being the team's chassis specialist, happens to be a cousin of Dale Jr.'s.

"They kinda hang out together," Eury Sr. says, "and they trust each other a lot, which is good."

At Daytona, their first race as a team, they qualified behind only Mike McLaughlin and Joe Nemechek, two superspeedway veterans. In the race, Dale Jr. ran with the leaders until he broke a drive shaft leaving the pits and lost several laps. Later, while logging miles to pad his experience, he got caught up in a backstretch tangle and flipped. Unfazed, he said, "We practiced good, qualified good, raced good, drafted good. But circumstances got us."

Circumstances couldn't get him every week. At Las Vegas, the kid ran a great 2nd to Jimmy Spencer; at Nashville he was a fighting 3rd; at tricky Darlington, a solid 10th; and at Bristol, he trailed only Elliott Sadler at the finish after starting from the pole.

Then came Texas, where he seemed to have lost any hope of winning when he stopped for tires late in the race, with Nemechek looking like the class of the field. But as Dale Jr. stormed back, every move he made suggested that he would be heard from before the checkered flag fell. And he was: when he caught Nemechek on the next-to-last lap and closed right up on the leader's bumper, using one of his dad's aerodynamic ploys to loosen up Nemechek's

car, it was clear that Junior had decided he wasn't about to run second.

"I want to be my own person and make a name for myself, but there's no pressure to be as successful as my father," —Earnhardt Jr. *Nigel Kinrade*

"We came off [turn] two, and we gained on Nemechek a little bit. I could see his car was out of shape; he wasn't very comfortable. Then, off four, we got a run on him, and I knew we'd be able to make a pass on the front straightaway.

"I went to the inside and got by him before we got to the corner, so we could get into the corner good and straight. I knew then that if I held my line and ran as hard as I could, we had a win."

Strapped in and ready to do battle on Bristol's tough half-mile.
Harold Hinson

His father was in victory lane even before Dale Jr. climbed out of the car, and their initial exchange—Dad leaning into the cockpit for a private word, the son tousling the old man's hair—will play on videotape as long as Junior races.

I ask Dale Earnhardt about that, and about the first words he said to his son. I expect him to either rerun the moment with great emotion, or tell me it is none of my business. Instead, he shrugs. "Just 'Good job, son. Great race. Congratulations.' Something like that."

Turns out Dale Jr. doesn't know for sure, either. "You know, I couldn't really hear him at the time, for all that was going on. But I knew he was happy."

Now things have been pretty much happy ever since. Sure, Junior has made mistakes—including those two straight wrecks his father complains about—but he has run more good races than bad ones. He has listened to Eury, listened to his dad, listened to Steve Park ("He's been really good about giving me advice," Dale Jr. says), and done a remarkable job of reading tracks he has never seen and competitors he has rarely faced.

Maybe the best example was the June race at Richmond. He lost the lead late in the going to Jeff Burton, who had the faster car, but then clung to Burton's tail and kept the outcome in doubt until the last lap. It was his first start at Richmond, and his first prolonged head-to-head scrap with a Winston Cup winner.

"Earnhardt did a super job," Burton said. "He ran me to death."

Tony Eury says, "We knew from last year that Dale Jr. was good. But I don't think we knew he was this good."

There is a chuckle here from Eury, who has now been down the road to stardom with two Dale Earnhardts. He says, "You know, Dale Jr. gets a little wild sometimes on the racetrack, and we've got to calm him down. But it's a lot easier to get that attitude out of a driver than it is to put it in one."

His father's legendary intensity showed in Earhardt Jr. at Charlotte in October of 1998. *Nigel Kinrade*

Meanwhile, Dale Jr.—that "competitive little driver"—continues to gain attention wherever he goes. Every time you turn on one of those weekly motorsports TV programs, some commentator is gushing over him. Every time you flip open the local newspaper on a Busch Series race day, you see his photo on the lead sports page. He is young, he is fast, and he is Dale Earnhardt Jr.

I ask him if he thinks he will ever be fully out of the shadow cast by his own name, and Junior frowns.

"Maybe one day, when Dad retires," he says quietly. "Maybe."

Now he is back to preaching patience, restraint. He says it is a mistake for people to expect him "to be awesome right out of the box." Even his famous father, he points out, needed time to become the Dale Earnhardt we know today. "They forget about the years when he struggled, when he didn't win many races, and he wasn't considered the best."

But that was then, this is now. There is no way to compare their situations, because so much is different: Dale Earnhardt had room to grow in those struggling years because his father, while a great champion, lived and died when NASCAR was essentially a Southern sporting phenomenon; Dale Earnhardt Jr., on the other hand, showed up as the son of a mainstream American icon. Like Hank Williams Jr., he was a household name before he knew it. It would be nice to think that stock car racing might grant young Dale some breathing space, but it just isn't going to happen.

He talks of the responsibilities his father has, the sponsor commitments and the businesses and the other demands, and he says, "It would be tough to be in his shoes. That's not something I'm looking forward to."

I mention this comment to Dale Sr., who says with a laugh, "It's going to happen. As your career grows, as it builds, all that stuff comes with the territory. It won't be a thing where he can say, 'I don't want to do this.' Because there's no way you can't."

"So you think he'll grow into those responsibilities," I ask, "the same way you did?"

Cameramen, reporters, or requests for interviews were never far away during Earnhardt Jr.'s rookie season in the Busch Series. *Stock Car Racing Archives*

Earnhardt the father says, "He'll have to, if he's going to make it in this business."

Maybe it is happening without the kid even seeing it. When my time with him is done, I step out of his team's transporter and recognize a couple of daily writers standing around, waiting their turn. Without even knowing it, I catch myself thinking, he is growing into those shoes.

Dale Earnhardt likes to say of his boy, "In a year or two, once when we get some time on him, he'll be OK."

Look again, Dad. Junior is OK right now.

4

LIVING UP TO THE HYPE

BY BRUCE MARTIN
From *Stock Car Racing*, August 1999

An Almost Cultlike Following Has High Expectations for Little E

Dale Earnhardt Jr. was focused on winning another Busch Series title in 1999, but there was another goal to be reckoned with: the five Winston Cup events his team planned as a prelude to its Cup rookie season in 2000. The buildup began early in 1999, with tracks at Charlotte, New Hampshire, Michigan, Richmond, and Atlanta chosen as the first five venues in his Cup career.

No rookie driver in NASCAR Winston Cup history has entered his first race with as many expectations, or as much hype, as Dale Earnhardt Jr. The attention that has been placed on the 24-year-old son of Dale Earnhardt is far greater than that for the late Davey Allison, or even for Jeff Gordon. And compared to the buildup for young Earnhardt's first race, Tony Stewart's debut in the Daytona 500 may have well happened in relative obscurity.

But is there any way that Dale Earnhardt Jr. can ever live up to that much hype?

"This is just rolling along and snowballing and we're going to see how far down the hill is," Earnhardt Jr. says. "I hope it keeps going for a long time. I didn't expect it to happen, but it's happening.

Being a NASCAR star means making appearances of all types, including making sales pitches. *Nigel Kinrade*

"I don't know whether to slow it down, to stop it for a minute, to take a break, to keep going, to speed it up. I don't really know because I don't have any experience with this. My father seems to think that more is better or more demand is better. The dogs are howling at the gate and he keeps on letting them in. Anytime anybody beats on the door, he keeps letting them in. That's cool, I guess, if that's the way it needs to be.

The red Budweiser Chevrolet was about to become a fixture in the Winston Cup garage. *Harold Hinson*

"You feel like everybody's looking at you. You feel like all eyes are on you."

Armed with a lucrative sponsorship from Budweiser, a famous father who is also his team owner, and the 1998 Busch Series championship earned in his first full season, the unassuming Earnhardt has created an almost cultlike following.

"The swarms are fun," Earnhardt says. "It's like at Atlanta, trying to go to driver's introduction last year. Those times are hilarious. They're great. I feel like a rock star every once in a while. It's kind of fun. At least you get to experience what they go through sometimes."

Son of The Intimidator

"I wish everybody could see me now," says the junior Earnhardt.

Especially those he went to high school with. As a youngster, his father shipped him and his sister Kelley off to Oak Ridge Military Academy near Greensboro, North Carolina. Young Dale learned a great deal of discipline before he returned to public school at Mooresville High School, but being the son of a famous father often had its drawbacks.

As he began his transition into Winston Cup, Dale Earnhardt Jr. found himself at the center of attention. *Harold Hinson*

"I had an interesting time going through high school," Earnhardt recalls. "I was kind of a loner during that period of my life. I was driving race cars and I guess that wasn't cool to a lot of the other high school kids. Being Dale Earnhardt's son wasn't an advantage, I'll tell you that. It was a disadvantage the majority of times. If he ran good, nobody ever said anything to you, but if he ran bad, you got picked on.

"I didn't understand why my father was booed so much, but I felt like at the time he was standing up for what he felt was right and that's real important.

"When I was in school, there were jealous kids. There were kids that never said anything to you when your dad won a race, but when he crashed out, they were the first ones to say, 'So, your daddy wrecked!' You learn to deal with that at an early age."

It was also while young Earnhardt was away at military school that his father was becoming the biggest name in NASCAR Winston Cup racing. This was at the peak of Earnhardt's career, when he intimidated opponents and won Winston Cup titles on a regular basis.

During Earnhardt's dominance, the sport enjoyed a tremendous spurt in growth, which kept Earnhardt away from home for sponsorship appearances and other demands on his time. Because of that, Dale Jr. was unable to spend a great deal of time with his father.

"The fact my dad wasn't with me when I was growing up, that's all right," Earnhardt says. "Who wants to be around his dad all day, every day? How many butt whippings can one person stand in one week?"

Earnhardt Jr. kept pushing his Busch efforts in 1999 with the knowledge that five Winston Cup races loomed large in his future. *Paul Melhado*

Media Star

Dale Earnhardt Jr. is a breath of fresh air in the Winston Cup garage area. Unlike Gordon and Stewart, who have come from outside of NASCAR's grass roots to become young heroes, Earnhardt appeals to the people who rooted for his father and at the same time portrays an image that major corporations can turn into an icon.

At only 24 years of age, he's almost too naive to realize how much he is in demand, but he does his best to give interviews, answer questions, and chat with the media. And when he talks, he sounds like Boomhauer from the TV cartoon *King of the Hill*.

All of this success came as a huge surprise to Earnhardt's grandmother, Martha. Despite the tremendous success of her late husband, Ralph Earnhardt, and her son, Dale Sr., she didn't expect the same from grandson Dale Jr.

"I haven't talked to her much about it," Dale Jr. says. "She was just surprised I was a race car driver. She didn't think I was going to make one. She didn't have much faith in me. I guess I don't look like I can get out of the rain."

Hard to believe, however, is that young Earnhardt can still enjoy a certain degree of anonymity, simply because the Busch Series drivers aren't as readily recognized as the Winston Cup stars. But Earnhardt also understands it's merely a matter of time before the fans will be able to pick him out of a crowd.

"I can pretty much still do whatever I want, whenever I want, when they give me the time to do it," Earnhardt says. "I think I get a half-day a week now. I'm already complaining about the time I don't have. Isn't that bad?

"I don't see why my dad doesn't like all that attention. You can be a regular guy at the house by yourself. When I was a little kid and I watched people give a race driver attention, I was like, 'Cool, man. That's awesome.' That's always the way I was. I think some people are like that and some people aren't. When people come up to me and say, 'You're one of my favorites,' I'm kind of shocked and surprised because I don't call myself a great race car driver yet."

Countdown to E Day

During his father's dominance, young Earnhardt always watched his father race at Charlotte Motor Speedway; back in those days racetracks were named after the towns they were located near, not by the company that paid the most money for naming rights.

Solitary moments at a racetrack, like this one at California Speedway, were few. *Jon Fitzsimmons*

Dale Earnhardt Jr. in his Winston Cup racing suit.
Harold Hinson

"We had a little area on top of one of the turns in the infield where our family parked, my dad's brothers and our own little group," Dale Jr. recalls. "We'd pitch a few tents and roll our little models and cars down the hill of the road course during the race. We just had a lot of fun. I remember doing that all the time."

As a youngster, Dale Jr. often dreamed that it was him in the race car as he rolled the cars down homemade speedways carved out of the infield at Charlotte Motor Speedway while his father did the real thing a few hundred yards away on the asphalt surface of the racetrack. That's why it is only fitting that young Earnhardt's first attempt at a NASCAR Winston Cup race was set for the Coca-Cola 600 on May 30. To capitalize on his dramatic debut, the team and sponsor even started a "Countdown to E Day" promotion. At one point, Dale Jr. began to call it "Countdown to Hell Day."

Talk about a lot of pressure to make the field for a NASCAR Winston Cup debut.

"I don't think there will be much pressure," Earnhardt says. "I'm going to be excited because I'll be at a Cup race for one. A real, legitimate Cup race, not a make-believe Japan Cup race. It will be a real Cup race where I'll be racing against the best and we'll truly see what I've got. If I have a car that can do that, we'll be there.

"We have the outlook of just taking the bull by the horns and we have a great opportunity to reap the benefits of all the hard work. For years and years, I've worked on trying to be a good race car driver. We had an excellent year last year and we sold a lot of souvenirs. We got a lot of fans interested and that helped my fans a lot.

"You just don't see that kind of reaction to a Busch driver."

Earnhardt got a sneak peek at Winston Cup competition last November when he competed in an exhibition race at the Twin Ring Motegi oval in Japan. It was a race that was highlighted when he ran into the back of his father's car a few times toward the end of the race and eventually passed him to finish sixth—two positions ahead of the seven-time Winston Cup champion.

Rather than beaming proudly and saying, "That's my boy!" the elder Earnhardt was red-faced and thought his namesake should respect his elders.

"Building up to the Japan race, he started considering me more of his competition, like just another guy for him to outrun," Dale Jr. says. "Around that area, he just started spending more time doing other things. I haven't seen him or talked to him much ever since.

Anticipation for Earnhardt Jr.'s move into Winston Cup was growing. *Nigel Kinrade*

The defending Busch Series champion and the defending Winston Cup champion, Jeff Gordon, spent time together at Lowe's Motor Speedway in 1999. *Nigel Kinrade*

Following 1999's final race, at Homestead on November 14, Earnhardt Jr. celebrated his second Busch title and displayed the trophies from both championship seasons. *Nigel Kinrade*

"It felt good to beat my dad at Japan because he didn't think I could do it. He always doesn't think I can do it and I like to prove it to him. When you do succeed, he still says, 'I could have done it better.' Or, 'No, you didn't do what you think you did. You're not as big as you think [you] are.'

"He always has some way of putting you down. That makes me keep crawling to get to the top. As long as he keeps acting like he acts, I'll keep acting like I act."

Perhaps it's Earnhardt's way of showing his son "tough love," but he has taken an almost critical approach to how he will treat his son on the racetrack.

"I'll race him just as hard as anyone else," Earnhardt says. "I'm not going to take it easy on him. But hopefully, in the end, it will be Earnhardt and Earnhardt. I'm racing for Richard Childress Racing and racing to win. Whoever else is out there, I'm racing to beat them.

"He has to get in and get started first. He's just a Busch Grand National driver now. We're not worried about him until he gets there."

Rather than be hurt at the way his father responds, Dale Jr. realizes he's being taught a lesson.

"He does make a point of making it as difficult and as much of a struggle as possible," the young Earnhardt says. "He goes out of his way to make things more difficult because he believes, in the long run, I'll be a better person for it.

"I think he is the best ever, but I can sort of understand how he's feeling. You don't want to be outrun by somebody you raised and who is not supposed to know what he's doing or be competitive.

"He's never been one to compliment me often. It's not that he never does, it just makes you insist on always trying to impress him. When you win a race, you do it also for your father, to make him happy because that's something he did a lot. When you do something that he did and not really match his success, but make a leap toward it, that kind of means a lot. Until you succeed or do

Earnhardt Jr. filled many autograph requests in 1999. *Nigel Kinrade*

something that he has done or something he's proud of, you never feel up to par. You never feel like you're playing with a full deck.

"Hopefully, I'll win the Daytona 500 before my 20th try. I'm not out to match his accomplishments or anything like that. I just want to be competitive, content, and satisfied with what I'm doing. I'm excited, proud, surprised. You can write the list and check them off."

High Expectations

Earnhardt began his professional driving career at the age of 17, competing in the Street Stock division at Concord Speedway in Concord, North Carolina. Within two seasons, Earnhardt had perfected his driving abilities and joined the NASCAR Late Model Stock division. He ran his first NASCAR Busch Grand National Series event in 1996 at Myrtle Beach, South Carolina. In the Busch Series last season, Earnhardt won more races, led more laps, led more miles, won more money and more poles than any other driver on the circuit this year.

Don Hawk, president of Dale Earnhardt, Inc., and one of the top businessmen in NASCAR Winston Cup racing, negotiated the Budweiser deal. Hawk says the team had high expectations for Earnhardt's first year in Busch Grand National racing, but admits the young driver exceeded those expectations.

"He went a little further than we thought, so that escalated the hype on him," Hawk says. "What we're doing in the Busch Series is what the 24 car [Jeff Gordon] has done in the Winston Cup series and we are moving the bar to the next level.

"He's only 24. The first time his daddy bent a steering wheel in Winston Cup, he was 27. Jeff Gordon has just turned 27. The key on Dale Jr. is he has a contract for five Winston Cup races next year, and five more years on top of that. When that contract is over with, he'll be one year older than his daddy was when he first bent a steering wheel and only one year older than Jeff Gordon is now.

"We're sitting on a 24-year-old keg of dynamite," says Hawk.

With so much happening so soon in his career, how will Dale Earnhardt Jr. keep from being overwhelmed from all the attention when he moves up to Winston Cup?

The September 11, 1999 night race at Richmond was one of the five Winston Cup events Earnhardt Jr. ran that season. He finished 10th after winning Richmond's Busch race the night before. *Nigel Kinrade*

What's better than winning a championship? Winning a second one ranks high on the list. *Nigel Kinrade*

Dale Earnhardt Sr. and wife Teresa helped celebrate the second Busch Series title, secured at Phoenix in the penultimate race of the season. *Nigel Kinrade*

"I've kept a level head through all of this and that comes from living in a double-wide, driving dirty pickup trucks," Earnhardt says. "That's what it's all about.

"Now, maybe Daddy will raise my allowance. I'm just going to be myself and do what I want to do. We don't want to get overwhelmed by all of this. We're lucky, very lucky, and we're blessed. Being in racing with my dad has brought us closer together because it's allowed us to spend more time with each other. That's pretty cool."

Gordon vs. Junior

Having one of the greatest drivers in racing history as his father is perhaps preparing Dale Jr. for an even bigger nemesis in the future—Jeff Gordon. This is the battle that race fans and the media have been looking forward to, although Tony Stewart may be in a position to crash the party with his fantastic rookie season. The Gordon foes are looking for a hero that can knock Gordon off his throne as the top driver in racing today.

Racing fans' hopes have created even more expectations for young Earnhardt.

At Darlington in March, Earnhardt Jr. grabbed a moment to himself. *Nigel Kinrade*

The media blitz was on with Earnhardt Jr.'s Winston Cup debut at Charlotte in May of 1999. *Nigel Kinrade*

A win at Dover in June 1999 kicked off a celebration.
Nigel Kinrade

"I'm ready to take on Jeff and everybody else out there," Earnhardt says. "I've actually had a lot of people talking to me about racing Jeff Gordon. For anybody even thinking I might be able to compete with Jeff Gordon is a compliment because he's the best out there at this moment.

"Me and Jeff have talked about it briefly. It's going to fill the seats a little bit if he has somebody racing him hard and doing it every week and battling for the championship."

Rather than flinch from the challenge, Gordon welcomes the competition that Dale Earnhardt Jr. will provide. Perhaps he needs that to stay motivated in a sport he has dominated unlike any other driver in the sport's history at such an early age.

"I think we all respect him," Gordon says. "He's got a lot of talent. I think just gaining the experience in the heavier Winston Cup cars is the only thing I see he's lacking in. But from what he's done in the Busch Grand National cars, that's not going to take him long to do in the Winston Cup cars.

The official celebration for the team's second straight Busch title came at Homestead in the tour's final race of the season. *Nigel Kinrade*

"There's no doubt, there's always going to be somebody younger, more talented. I've been saying for a long time, they are out there. It's just some of these car owners need to look a little bit harder or take a chance, like Rick Hendrick did with me. Now we're starting to see some real talent stepping up that everybody sees, like Dale Earnhardt Jr. and Matt Kenseth. There are a lot of guys all over the country that we don't get to see that have far more talent than those guys. They just don't get the opportunity to show it.

"Dale Earnhardt Jr. is doing an unbelievable job. I look forward to seeing him come in. He'll be successful because Dale will make sure he is in good equipment, safe equipment, and that is an important part of it. And he's brought him up right. There are some things that I missed out on that you can't do over again, and that's learning more about a stock car before I was able to get to Winston Cup. For me, I learned all about midget and sprint cars, but all of a sudden, I'm in a stock car and I'm learning how to drive it. They were able to learn the stock cars before they ever got to Winston Cup."

Rock Star

Earnhardt is so young, so new to all of this, that he often does a double take when he sees a race fan wearing a T-shirt with his picture on it and cheering him on.

A second successful Busch season firmly established Earnhardt Jr. as a NASCAR star. *Nigel Kinrade*

Earnhardt Jr. and the ACDelco team looked super again in 1999. *Nigel Kinrade*

Dad's International Race Of Champions victory at Daytona in February brought a plain-clothes visit to victory lane. *Nigel Kinrade*

The Bud Chevrolet looks low and lean during the team's second Cup appearance of the season in July at Loudon. *Nigel Kinrade*

How do you celebrate a Busch Series title? With Busch beer, of course. *Nigel Kinrade*

Earnhardt Jr. looked spiffy in his new Bud duds at Michigan in August. *Nigel Kinrade*

As defending Busch champion, Earnhardt Jr. earned a spot in the 1999 IROC series. *Nigel Kinrade*

Bill France (second from right) and Dale Earnhardt Sr. shared the spotlight at Homestead during the celebration for the Busch championship. *Nigel Kinrade*

"It feels good, but there's never enough, though—you always want more," Earnhardt Jr. says. "There's always a race fan walking by with a Rusty Wallace T-shirt on and you wish he had yours on.

"By being myself and doing the best I can on the track, race fans pay a lot of attention to your personality because that's something that's hard to see. You have to pay attention to what you do on and off the racetrack. If you do the right things, and say the right things, and try to be true and honest to the race fan, he's going to go over there and grab your hat off the shelf. That's what I try to do.

"I can't grow up any faster than I am right now. I'm finding out a lot of new things about myself. Yes, I'm still young."

His youth and inexperience are reasons why he believes he will have no trouble adapting to the Winston Cup Series after being an instant hit in the Busch Series.

"I expect to qualify for the five races and not have to go home from them," Earnhardt says. "I expect to be able to lead a couple of them. I expect to be competitive in all of them.

"The fact that I'm young is probably an advantage. I probably have no clue or idea what's really going on. Somebody told me after we won the Busch championship that it would be 10 years down the road before I really realized what I had accomplished.

"I don't think there's going to be much of a transition. The hard part is going to be juggling the two schedules this year. Even with five races, it's going to be tough on the crew and me. I went to Japan and drove a Winston Cup car. It might not have had the most horsepower or it might not have been the best race car, but I didn't see much difference.

Earnhardt Jr. won the Busch race at Michigan in August of 1999, besting Winston Cup regulars Jeff Gordon, Mark Martin, Jeff Burton, and Michael Waltrip. *Nigel Kinrade*

"I want to win. That's what we came here for. We've probably got a long way to go and a lot of things to figure out before we're at the level of Jeff Gordon and a lot of these other guys. And I understand that."

With so much hype and anticipation for Earnhardt's arrival into NASCAR Winston Cup, he has already been called the "Face of NASCAR in the 21st Century." Rather than shy away from such lofty expectations, Earnhardt looks forward to fulfilling the promise.

"Hopefully, when they write that book or that magazine and they write about the 50 best drivers, 50 years from now, I'll be on that list somewhere because I hate that I missed this last one," Earnhardt says, referring to last year's 50th Anniversary celebration. "I guess I was just too young."

5

ROOKIE E

Lessons Learned

From *Stock Car Racing* magazine, July 2000,

By the end of 1999, Dale Earnhardt Jr. had won his second straight Busch Series title and was looking ahead to his inaugural Winston Cup campaign. He compiled 13 victories during his two seasons running full-time in Busch and became one of the top young drivers in NASCAR. Then it was on to bigger and better things in Winston Cup, where Earnhardt Jr. had his debut run at Charlotte in May of 1999. His best finish in five Cup races that year was a 10th at Richmond. He entered the series full-time the next year as part of one of the best rookie classes in Winston Cup history.

The 2000 Winston Cup season started with what was considered by many to be the most competitive rookie crop in at least two decades, maybe ever. Racing observers frequently compared the rookies in the class of 2000 with that of the 1979 campaign, which included seven-time Winston Cup champion Dale Earnhardt and Harry Gant.

"I think it's one of the best and most diverse as well," says NASCAR historian Bob Latford. "There's so much talent—proven talent—whether in Busch, or in CART, or wherever they came from. It's going to be a tough one and probably one of the best groups in a long time."

The list of first-year drivers includes Dale Earnhardt Jr., Matt Kenseth, Stacy Compton, Scott Pruett, Ed Berrier, Dave Blaney, Jeff Fuller, and Mike Bliss.

Shades of the late-model days and Earnhardt Jr. working on his car still flashed during the rookie campaign. *Nigel Kinrade*

A win at Texas in April came in just the seventh race of Earnhardt Jr.'s rookie season in Winston Cup. *Nigel Kinrade*

Yet, despite their experience in other series—some of it considerable and noteworthy—making the move to Winston Cup is a whole other animal.

For some, such as Earnhardt Jr. and Kenseth, it means going from the top of the Busch Series ladder to the bottom of Winston Cup. And for guys like Pruett, who laughs at the fact that he's a rookie again at 40, it means learning a new way of driving.

This year's rookies also have to follow the success of last year's rookie winner, Tony Stewart, who had the best season ever for a rookie driver, earning three victories and finishing in the top ten in points. In one season, Stewart set the bar higher for all others.

Ratcheting up the scrutiny of the rookie drivers is the arrival of Kenseth and Earnhardt on the Winston Cup Series. The two battled for the Busch Series title and their teams draw a lot of attention to the rookie race.

Living up to Stewart's record is not something Kenseth says he worries about.

"I don't think it's realistic to expect that of any of us," says Kenseth. "He had an exceptional series. I think Tony Stewart is the exception, not the norm. The best we can do is the best we can do. Every week we need to do the best we can do. If the best we can do is 13th place, fine."

Dale Jr. and the Eurys, Tony Sr. and Tony Jr., celebrated at the track that also brought their first Busch win in 1998. *Nigel Kinrade*

A pair of Winston Cup rookies, Tony Eury Jr. and Earnhardt Jr., talk things over at Pocono in June of 2000.
Nigel Kinrade

Dad checks things out before Earnhardt Jr. hits the track.
Harold Hinson

In 2000, Earnhardt Jr. was a rookie sensation in Winston Cup, and he was just three years removed from his late-model career. *Nigel Kinrade*

Earnhardt, who must also live in his father's shadow, has said repeatedly he's not going to let outside influences put more pressure on him than he's already putting on himself.

The struggle to make races only adds pressure to the already tension-filled world of being a Winston Cup driver. Yet none of them are going to let the added pressure of being rookies and the glare of the media spotlight get the best of them.

"The best thing that probably happened to me was when Gary Nelson sat down with me and told me some things that nobody else had said to me before," says Earnhardt Jr. "Even though a lot of good things have happened to me in my career, I sometimes need to be told that I'm a good race car driver. Just like being in a marriage. Even though your wife knows you love her, you've still got to tell her sometimes. That was kind of the situation I was in. It was like it'd be nice to get patted on the back sometimes even though we had a bad weekend. To hear it from somebody like Gary Nelson, who can't play favorites and to have nothing to gain by telling me that I'm going to make it, that sounds pretty good to me."

Earnhardt Jr. says he learned other lessons early in his rookie season, particularly on restrictor plate tracks. *(continued on page 87)*

THE FIRST VICTORY

At Texas Motor Speedway in 2000, Dale Earnhardt Jr. did a lot to improve the image of rookies by winning his first Winston Cup race in only the seventh event of the season.

Did he think he could win in his rookie year?

"I thought so," he said from victory lane. "I think we're going to see a couple of rookies get into victory lane this year. I didn't know how quickly it would come. I'm kind of overwhelmed with what happened today. It's just crazy, man.

"After the last three weeks, it was fun to get out front and show these guys I could use my head and make smart decisions. Even when I've got a real fast car and I'm not up front, I can be patient and pick my way through there. It was fun to prove that to them and prove my status as far as a driver goes."

Four races later, at Richmond International Raceway, Earnhardt Jr. claimed his second Winston Cup victory. His momentum continued in the tour's next event, at Lowe's Motor Speedway, as he won the pole for the Coca-Cola 600. The race, however, belonged to fellow rookie Matt Kenseth, who won for the first and only time in 2000. Kenseth also edged Earnhardt Jr. for 2000 Winston Cup Rookie of the Year.

Giving dad a love tap at Talladega. *Harold Hinson*

Elliott Sadler was fresh off a runner-up campaign to Rookie of the Year Tony Stewart when he and rookie Earnhardt Jr. talked at Daytona in February of 2000. *Nigel Kinrade*

After battling for two seasons in the Busch Series, where Earnhardt Jr. won back-to-back titles, Matt Kenseth and Earnhardt Jr. brought their friendly rivalry to Winston Cup in 2000. *Nigel Kinrade*

In Busch, Kenseth had a runner-up points finish in 1998 and a third-place effort in 1999. *Nigel Kinrade*

Kenseth gained the upper hand by edging out Earnhardt Jr. and winning Rookie of the Year in Winston Cup. *Nigel Kinrade*

Earnhardt Jr. displayed his short-track prowess with a win on Richmond's three-quarter-mile layout. *Nigel Kinrade*

(continued from page 83)

"When you first come in as a rookie, it's really hard to know when to go, when to make a run, when to attempt a pass, and when you need to stay in the draft," he says. "It was real hard for me when I first started running at Daytona and Talladega to know when to do that, and when not to, and what was a smart move.

"The whole time you're out there, especially when you're a rookie, all you're trying to do is

Rookie E sits for a studio photograph at Daytona in February. *Nigel Kinrade*

The night race at Richmond on May 6, 2000, became the scene of Earnhardt Jr.'s second win, tying him with Davey Allison for the fastest two wins in Winston Cup history. Both men won their 2nd race in their 16th attempt. *Nigel Kinrade*

impress the guys around you. You've watched Terry Labonte and Bill Elliott and all those guys race for years and all you're trying to do is to do things to impress them, to where you'll get out of the car and they'll pat you on the back. I mean, that's all you're looking for.

"So it took me awhile, took me a couple of races, a couple of years. Like in the Busch Series, I crashed out pretty much every one of the plate races I ran. And that was just because I didn't know when to be aggressive and when not to be aggressive and just kept getting myself in the wrong situations.

"Then when you come into the Cup series, you've got to know when to be in the low line and you've got to know when to be in the high line. You've got to know when changing lines is going to improve your position. It's real tough and it's hard to figure that out. I used to race as hard as I

FORGET DYNASTY—WE WANT A RIVAVLRY

BY JEFF HUNEYCUTT
From *Circle Track* magazine, October 2000

Richard Petty had David Pearson, Darrell Waltrip had Cale Yarborough, Superman had Lex Luthor. For a while, poor Jeff Gordon had nobody to keep him from rolling to double-digit victories, which made for a couple of pretty boring Winston Cup seasons. The best rivalries, of course, are the ones that are between competitors or teams battling over championships. Think Lakers versus Celtics, McEnroe versus Connors, Godzilla versus that bug thing.

On the Winston Cup side, we have the makings of a fantastic rivalry bubbling up right in front of us. Dale Earnhardt Jr. and Matt Kenseth, both Cup rookies, have already done something in the first half of the season that many veterans in the field can only dream of: They've won races. In fact, Earnhardt Jr. has already won three including The Winston. Kenseth finished an amazing sixth in his first-ever Winston Cup race in 1998. Both have the makings of future greats—and rivals.

One of the keys to growing a great rivalry is that the combatants keep bumping into each other. Earnhardt and Kenseth do seem to keep squaring off against one another. In two seasons on the Busch Series (1998–1999), Earnhardt claimed the championship, and Kenseth was his constant and greatest challenger. Although Earnhardt has claimed most of the limelight with 2 wins after 14 races, he didn't lead the rookie standings thanks to Kenseth's steady string of solid finishes.

Interestingly, despite their battles on the track, there is no animosity between the two. In fact, they are honestly friends. When Kenseth won his first Cup race in the Coca-Cola 600, Earnhardt—who had dominated the early stages—was one of the first to congratulate him. Both respect the other as a driver although their styles differ greatly. Kenseth is as cool as a cucumber, while Earnhardt is a hard driver who seems to have inherited his competitive fire from his daddy, seven-time Winston Cup champ Dale Earnhardt Sr. So here's to hoping that in the not-so-distant future, one of these young men will be challenging for his first championship, and the other will be right there, trying to take it away.

The Earnhardts were tight at Daytona for the season-opening IROC race in 2000.
Kevin Thorne

could and never get anywhere, almost lose three or four spots. And my dad could whip right up through there, no problem. I would be like, 'Man, how come I keep getting in the wrong line and he seems to keep getting in the right line?' And you just kind of figure that out. There's no way to really tell you, 'All right, this is how you do it to get to the front.' You just have to be sitting there in the car when the time comes to know, 'All right, this line is getting ready to move forward three or four lanes. I'm getting in it.' You hop in it and pick up three or four spots and get back to the bottom line and make a few moves there. And you do it without really being aggressive. You hope that when you're done, you're up front and everybody wonders how you got there, you know? That way you haven't caused too much trouble doing it."

Earnhardt Jr. earned a spot in the IROC Series by winning the Busch championship. *Paul Melhado*

ORDINARY GUY, EXTRAORDINARY LIFE

6

BY THOMAS POPE
From *Stock Car Racing* magazine, January 2001

Trying to Remain a Regular Joe

As Dale Earnhardt Jr. tackled the challenges of being a Winston Cup star, he had every reason to become a changed person, intent on living the celebrity life and counting the money that was piling up. Much of the same Earnhardt Jr. remained, however, from the guy who ran late-models and hung out with the guys at Myrtle Beach. Through the whirlwind of attention and high expectations that came from being the son of Dale Earnhardt, "Little E" was striving to be himself.

Dale Earnhardt Jr., despite millions pouring in from all angles, has tried to remain a regular Joe and he keeps his racing career as his top priority. Earnhardt Jr. says the trick to staying that way is to remember where you came from. *Harold Hinson*

Dale Earnhardt Jr. is, in business parlance, multitasking. He's just stepped back into his Mooresville, North Carolina, home after paying a visit to Dale Earnhardt, Inc. There are four race shops there, totaling 200,000 square feet, but he's been to the biggest—the sparkling, sprawling main facility accurately christened the "Garage Mahal" by Darrell Waltrip. It's all just minutes from Dale Earnhardt Jr.'s house, which is located on the same 350-acre site. His father, the seven-time NASCAR Winston Cup champion in pursuit of an eighth, had summoned him over to take part in a press conference where it was announced that Dale Jr. and Steve Park will have another Cup teammate in 2001: Michael Waltrip.

The announcements made, the interviews out of the way, and the photos snapped, Earnhardt Jr. has made the short trip back to his sanctuary, pecking away on one of the four computers in his house. He's chatting long-distance to yet another reporter, and doing so without missing a beat while simultaneously checking his email and surfing the Internet. He's liable to run across an audio file he'll download and burn onto a CD, and it could be anything across the musical landscape—from Joe Cocker to Rush to Neil Young.

Despite enjoying a phenomenal year, including 2 wins in his first 16 races, Earnhardt Jr.'s first full season in Winston Cup had the usual rookie mishaps. *Harold Hinson*

Diamond to Elvis

In other words, he's just being 25-year-old Dale Earnhardt Jr., the hottest commodity in NASCAR racing for most of the last three years. At least one day a week, publicist Jade Gurss tries to keep Earnhardt Jr.'s schedule free of interviews. As willingly accommodating as Earnhardt Jr. is, six days a week of the same type of questions—"If you and your dad are racing for the win on the last lap, would you wreck him?"—is about all he can stand.

"I don't mind that much. I look at every interview as a chance to improve my speaking skills and learn what's tasteful and distasteful," he says. "There have been some things I've said and shouldn't have said. Everybody has to learn the ropes.

"Like the press conference today. I basically went over to show people I'm down with my father hiring Michael to run this third Winston Cup car. I know I must've gotten interviewed five or six times, and in every one of them somebody asked me, 'What do you think about Michael?' I don't want to give people the same answer over and over, so I looked at it as a challenge and tried to answer it differently every time and speak intelligently about it. I love doing stuff like that."

A Star Is Born

At the age of 17, Dale Earnhardt Jr. began his racing career by driving street stocks around Concord (North Carolina) Motorsport Park. Eight years later, he is a two-time champion of NASCAR's Busch Series; the winner of two Winston Cup races in his rookie season, as well as NASCAR's all-star race, The Winston; and has earned more than $5 million on the track in the past three seasons.

He's the adopted favorite of his father's fans and will surely inherit their full loyalty when "Big E" puts his driving career in park sometime this decade. Yet there's no shortage of fans who pepper the stands clad in the rich red colors of the number 8 Budweiser Chevrolet.

He's handling it—the on- and off-track expectations—with a maturity that is rare among 25-year-olds.

Earnhardt Jr. is gifted with his father's skill behind the wheel, including a win-or-wreck-trying attitude. *Nigel Kinrade*

"He surprised a lot of people, including me," Dale Earnhardt says. "We knew he could drive a race car, but until you start winning on the racetrack and see how you react to the daily pressure, it's hard to know what you've got inside you."

"You can harness pressure, and sometimes it can be your enemy. You've got to use it as a positive," Dale Jr. says. "When you go back to the house or the motorcoach, that's the time to forget about it. You've got to be strong with it.

"This year has been easier than two years in Busch [where] I stepped into a car that had won. If I was going to prove myself, it was going to be in that car. If I didn't win, I was a schmuck. This is a new team. So maybe we've got some excuses, if you will. Budweiser has been very good to me, giving me days off. I have a lot of time to myself. I don't want to get burned out. I want to stay in this sport for a while."

Based on the lessons he's learned as a Winston Cup rookie, he's on the right track.

The Winston Cup Road

As a whole, Winston Cup racing hasn't been very intimidating to the son of The Intimidator. It wasn't intimidating last year, either, when he and Budweiser scheduled five chances to make Cup races. He batted a thousand, making all five races in first-round qualifying and posting a top ten finish—a sixth—in Richmond, Virginia.

Although he isn't a big spender, "Little E" is determined to enjoy his good fortune and not let his growing wealth go to his head. *Sam Sharpe*

Earnhardt Jr.'s best runs early in 2001 came on the 1- to 1.5-mile tracks. *Paul Melhado*

There were a few bumps and bruises along the way as Earnhardt Jr. trudged through the second half of his inaugural Winston Cup campaign. *Nigel Kinrade*

He carried that qualifying momentum over into 2000, grabbing a top-ten starting spot in six of the first seven races, including a front-row berth at the 1.54-mile Atlanta Motor Speedway. The first six events of the season produced mixed results: a top ten and two top twenties, followed by three straight runs of 29th place or worse.

Race number 7 hit the racing world like a sledgehammer. Nine days after Steve Park gave Dale Earnhardt, Inc. its first Winston Cup pole position, Dale Earnhardt Jr. brought home the organization's first victory.

At Texas Motor Speedway, another high-banked 1.5-miler, Dale Earnhardt Jr. simply stomped the stuffing out of the best stock car drivers in the world. He led more of the DIRECTV 500 than anyone else, blitzing to victory lane on the same piece of Fort Worth real estate where he scored his initial Busch victory just two years earlier.

Crew chief Tony Eury and chassis specialist Tony Eury Jr. "built a good car, and I pretty much did what I wanted in the car," Earnhardt Jr. says. "It did what I asked it to do. We didn't change anything all day. When we got a set of tires, I just adjusted my driving to them. I'd just point and shoot, and that thing went.

"On the last lap, I was so excited, and I could see the fans getting excited. When I turned through Turn 4, I saw all the flashbulbs flashing. I couldn't believe it—people screaming and cheering. I couldn't believe it."

The suds had barely stopped flowing at "Club E," the romper room in the basement of his home, when he had reason to celebrate again. Four weeks later, at Richmond (Virginia) International

Earnhardt Jr. realized he stepped onto a stage unlike any other when he made the move up to Winston Cup. *Nigel Kinrade*

Fans swarm whenever Earnhardt Jr. shows up. *Nigel Kinrade*

Raceway, Earnhardt beat on Tony Stewart going down pit road, and then beat everybody else to the checkered flag. The collision on pit row left Stewart with a flat tire and raised hackles, but Earnhardt insisted he had no other course of action.

"I would have been pretty happy to finish second behind Tony," Earnhardt Jr. says. "Then we had the caution, and Tony kind of made it tight coming out of the pits, and I did all I could to keep from running over the boys in [Geoffrey Bodine's pits]. The guy who was changing the right rear tire on [Bodine's] car was jumping out of my way. I couldn't have gotten any closer without hitting that car.

"It was just a bad situation. Tony was coming out of the pits behind us, so he couldn't just make a hard right and left and come out of the pits. He came straight out of his box. I guess he just didn't anticipate me coming out at about the same time.

"The pit-road deal could have easily torn my tire up. I was lucky and fortunate it didn't, but it did tear up the fender bad to where the car didn't handle the same as it was. We were still good enough and had new tires where we could hang on there. We had a pretty decent lead. I don't think if I hadn't been able to get by my dad and get that lead we had that we'd have won the race. Terry [Labonte] and them were pretty strong there at the end."

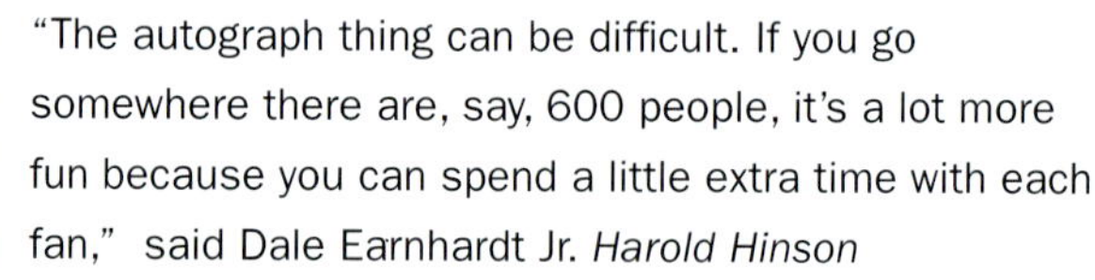

"The autograph thing can be difficult. If you go somewhere there are, say, 600 people, it's a lot more fun because you can spend a little extra time with each fan," said Dale Earnhardt Jr. *Harold Hinson*

Sixteen career starts with two wins is a feat only Davey Allison had also accomplished. But the spring rampage wasn't over yet. The Texas victory earned Earnhardt Jr. a place in The Winston, held at the high-banked, 1.5-mile Lowe's Motor Speedway. And just as he had done at Texas, Earnhardt Jr. almost toyed with the competition, becoming the first rookie to win the all-star event.

Four nights later, he returned to the same track to capture the pole position for the Coca-Cola 600. He hammered the field again, leading the most laps, but in the end, the car didn't handle as well as it had earlier in the night. As Earnhardt Jr. coasted back to the garage

Earnhardt Jr. managed some downtime at Dover in June of 2000—a rarity for the hottest rookie ever in NASCAR. *Nigel Kinrade*

There's no mistaking that Earnhardt Jr. has his own distinctive style, which is one more reason fans identify with him. *Harold Hinson*

area with a solid fourth-place finish, another rookie—Earnhardt Jr.'s good buddy, Matt Kenseth—rolled into victory lane.

In less than three months, Dale Earnhardt Jr. had more than lived up to the incredibly high expectations heaped on him. "I kind of blew 'em off there at the start of the year and didn't think much about it," he says of dealing with others' expectations. "We knew we had excuses to fall back on if we didn't run well because we were a rookie team. I mean, we had won two Busch Series

Earnhardt Jr. led the way at Michigan. *Nigel Kinrade*

Despite starting from the pole, Earnhardt Jr. finished 31st at Michigan on August 20, 2000, but he still looked very much like the son of The Intimidator. *Nigel Kinrade*

championships, but this was a completely different deal. We were all pretty much new to Winston Cup racing. The cars are different, and the engines make a bunch more horsepower. It's tougher to get the car balanced just right. So we had our excuses ready. Then we won a couple of races and I was thinking, 'This is cool. We're one of the teams to be reckoned with.'"

The Slump

Then came the summertime blues.

The Budweiser Chevys weren't bad, but they and their driver weren't running up front anymore. Earnhardt Jr. didn't lead a single lap for two full months, from mid-June until mid-August, when he won the pole position at Michigan Speedway and led 13 early circuits before winding up 31st, a lap down. The road courses—Sears Point Raceway in northern California and Watkins Glen International in upstate New York—proved particularly troublesome. Earnhardt Jr. tested at Sears Point and took several off-track detours trying to get the hang of its corners, and finished a middling 24th when the circuit went there to race a month later. At Watkins Glen, he tore up his primary car in practice and crumpled the backup in the race.

"We were all disappointed," he says. "We changed some personnel around, and we put pressure on ourselves to get back to victory lane.

"When I ran late-models, I didn't win much, but I didn't care—it was just a hobby. I was just having fun traveling up and down the road, and my buddies and I were all together. Then when I started Busch racing, my thought was, 'It's time to get serious. Losing won't be accepted.' But I really didn't have to deal with the downtimes then because we were winning races and championships, and we were always the focus. You know, everybody's saying it's one of the best things to happen in the Busch Series in a long time, [the rivalry between] Matt and me. TV ratings for the Busch Series were up.

"We moved to Winston Cup and thought we could do it, and we were all excited about it. It's weird how some of this stuff has turned out. We ran decent, but we struggled

At Pocono, Earnhardt Jr. and Dale Jarrett chat during driver introductions. *Nigel Kinrade*

at Rockingham and Darlington at the beginning of the year like I thought we would. Well, we went back to Darlington for the Southern 500, ready to fail, but did well and finished 11th. We tested at Martinsville in the spring and were really happy with the car, then went back and ran like crap, even though we did have battery problems. At Richmond, we win in May, then go back and test for the fall race since we were in the No Bull 5 deal, and we got the car even better in the test than when we had won there. We go back for the race and had a different right-side tire than we had tested with, that kind of threw our setup off, and we struggled in the race. I'm thinking, 'Damn, what the hell? We can't get the car to turn.'

"Things like that happening every four out of five weeks make you realize how tough a climb up the hill it is in Winston Cup racing."

That, perhaps, has been his biggest adjustment: realizing that he's not always going to have the best car, week in and week out. The Eurys are also learning about how to make magic with Winston Cup cars, but they don't accept any excuses.

"They feel very responsible for that car, even when there's nothing they could have done to keep mechanical stuff from going wrong," Earnhardt Jr. says. "When the battery goes dead, my reaction is to say, 'We would've had a great run if not for that battery.' They don't react like that. They get all pissed off, and that sends a message to the rest of the team about the fire and determination it takes to be successful in Winston Cup racing. When Tony Sr. gets that attitude, the guys

Under the watchful eye of Dad at Talladega during his rookie season, Earnhardt Jr. gave Rusty Wallace a friendly nudge. *Nigel Kinrade*

on the team know they're either with him or they go home. He's like Bobby Knight without the grabbing and cussing. He knows how to get his point across even though he's not an easy guy to approach. He's a very good motivator.

"We don't need a lot of motivation, though. When we won the first Busch championship, the motivation the next year was to do it again and prove that it wasn't a fluke. Then our motivation when we moved to Winston Cup was to be the first Busch team that's moved up to Cup and done well. Now our motivation is for people who don't think we're serious, to prove them wrong. But I don't have to rely on Tony or Tony Jr. to fire me up if I get slack. I search for things to motivate me. I get easily psyched."

Dale Earnhardt Jr.'s job doesn't begin and end with the race car. He's been heavily involved in marketing programs for Budweiser, but not more than he's comfortable handling. And then there is the ongoing, everyday business of dealing with fans.

Although earning the pole for the May race at Charlotte in 2000, Earnhardt Jr. finished fourth as fellow rookie Matt Kenseth grabbed his first Winston Cup victory. *Nigel Kinrade*

The Other Side of Life

In Mooresville, he says, he can still maintain some semblance of a normal life. It is, he notes, "Race City, U.S.A.," the home to more NASCAR teams than anyplace else, and fans aren't as fazed by a driver's presence in their midst. If he needs to stock up on groceries, toilet paper, or more Bud, he can do his own shopping without too many interruptions.

"Ninety-nine percent of the fans are great, but that other one percent—man, I don't know," he says. "There's always that one who's had too much to drink or wants to talk about his wife and how bad she wants to have sex with me. That's weird, man—my mind ain't wanting to try to understand that. Or the girl who wants you to sign her panties—that embarrasses me in public.

"And I go to races or an appearance someplace, and some of these fans have bags full of stuff they want me to sign. Where's it going? I'm hoping it's for family or friends, but there are two or three every weekend that are taking that stuff and selling it on eBay, and that's just not right.

"The autograph thing can be difficult. If you go somewhere there are, say, 600 people, it's a lot more fun because you can spend a little extra time with each fan. Like, 'Where'd you get this car?' Or, 'Boy, this is a nice diecast,'" he says. "When you do 2,000 people, you can't spend that kind of time with the fans because you can only sign 1,000, and you leave 1,000 pissed. You've got to hurry up. Of the 1,000 you sign, the guy in front's pissed because you ran him off, the one at the end of the line is happy because he got his autograph, and the ones in the middle don't know what to think because they got herded through like cattle. It's a hard situation—you definitely can't make everyone happy, but I try to do the right thing at the time."

And then there's the money—lots of it. Until his racing career took off, he figured he'd spend a good chunk of his life as a mechanic at Dale Earnhardt Chevrolet in Newton, North Carolina. He was doing that for $200 a week when he raced late-model stock cars and won only 3 times in 112 starts. Now that he can afford anything his heart desires, he describes himself as a relative tightwad, considering his net worth.

He's bought some cars—a 1969 Camaro, a 1971 Corvette, and a 1996 Impala that's loaded down with an earth-shattering audio system—and a pickup truck. He's got his house, a 25-foot

Earnhardt Jr. peered into his father's Chevrolet at Rockingham in October of 2000. *Nigel Kinrade*

At Charlotte for the running of The Winston in 2000, Earnhardt Jr. rode alone during pre-race ceremonies.
Nigel Kinrade

boat, his computers built primarily for video gaming, and "Club E." The rest of it?

"Of the money I've made since the beginning of 1998, I've still got 90 percent of it, if not more than that," he says. "Most of it is in stocks and money market funds. My dad introduced me to the right people, and they said, 'This is what you need to do.'

"When I was driving late-model stocks, I knew my dad was doing really well financially, and I started thinking, 'What if I do well? I want to be smart. I don't want to be broke at 50,'" he says.

"Let me tell you a story about that—about how I am with my money. When I was racing at Myrtle Beach, I started dating this girl down there, and everybody down there was driving big Chevy pickups, and I've got this little red S-10 pickup. I'm figuring if I'm going to impress her, I've got to have one of those like everybody else does.

"Well, I took the truck down to the dealer to have some brake work done, and I saw this big Chevy. Remember, I'm paying $100 a month for my truck, and the dealership's paying me $200 a week, but I went out and bought this big ol' truck, and I'm stuck with a $340-a-month payment. I drove that truck for about three months and realized, 'Man, this is too expensive. I don't give a damn whether this girl likes it or not—she's got to like me for me. I can't afford this.' I sold that truck for $16,000, paid off the $14,000 I owed on it, and bought a 1986 S-10 that was six years older than the one I had had to begin with. I drove that thing for four years and saved every dime I had."

The Intimidator's son became the first rookie to win The Winston, NASCAR's annual all-star event. *Kevin Thorne*

Earnhardt Jr. celebrated The Winston win in Charlotte's victory lane with Dad. *Harold Hinson*

The Earnhardts stand with other drivers included in the R.J. Reynolds bonus program at Talladega in April of 2000. *Nigel Kinrade*

Solving the complexities of Winston Cup racing are far more difficult than being fiscally responsible. But it's obvious only one year into Dale Earnhardt Jr.'s career that the pieces are all in place for a long, successful career—one that may have the components to match the best father-son team NASCAR has ever seen, Lee and Richard Petty. Earnhardt Jr. has youth on his side as well as his father's guts. He's got an eight-figure sponsorship to pay for the best equipment money can buy or build, and proven champions in the Eurys in his corner. As for himself, the challenges he's handled—dealing with success as well as defeat—prove that Dale Earnhardt Jr. not only has his feet on the ground, but also has his head screwed on straight. That's multi-tasking for you.

7

BEHIND THE WHEEL

BY DALE EARNHARDT JR.
From *Stock Car Racing* magazine, September 2001

Dale Jr. Writes About His Experiences Behind the Wheel of the Budweiser Chevrolet

It started out as a simple question at a midweek press conference prior to the 2001 Coca-Cola 600 at Lowe's Motor Speedway. "What do you see and feel out there?" a reporter asked Dale Earnhardt Jr. The question and Little E's subsequent answer clicked with David Bourne, editor of *Stock Car Racing*. Later, at Bourne's request, Earnhardt Jr. agreed to pen a cover story for the magazine.

Junior took us inside the cockpit of his number 8 Budweiser Chevrolet, offering rare insight into a 200-mile-per-hour sport. Racing With "E," complimented by Doug Miller's excellent portrait photography, became one of *Stock Car Racing*'s most popular stories in 2001.

Stock Car Racing magazine asked me if I could offer some of my thoughts about what it's like inside the cockpit of a Winston Cup car while we're racing for 400, 500, and even 600 miles. I'll try to describe some of the sensations and what it feels like to me.

First of all, driving a race car for a living is fun. It's fun as hell driving that fast whether we're racing or practicing. On the other hand, I'd rather not have to test so much because it's not so exciting. You're out there trying to be as much of a robot as possible so the team can make a real comparison between the changes they have made to the car. I have to be as accurate and as consistent as I can be so we can find out what works and what doesn't. It can be challenging, but most of

By time this photograph was taken, Earnhardt Jr. was one of the most recognizable drivers in auto racing.
Doug Miller

the time during a test, I'm wondering how I could have a CD player installed in the car.

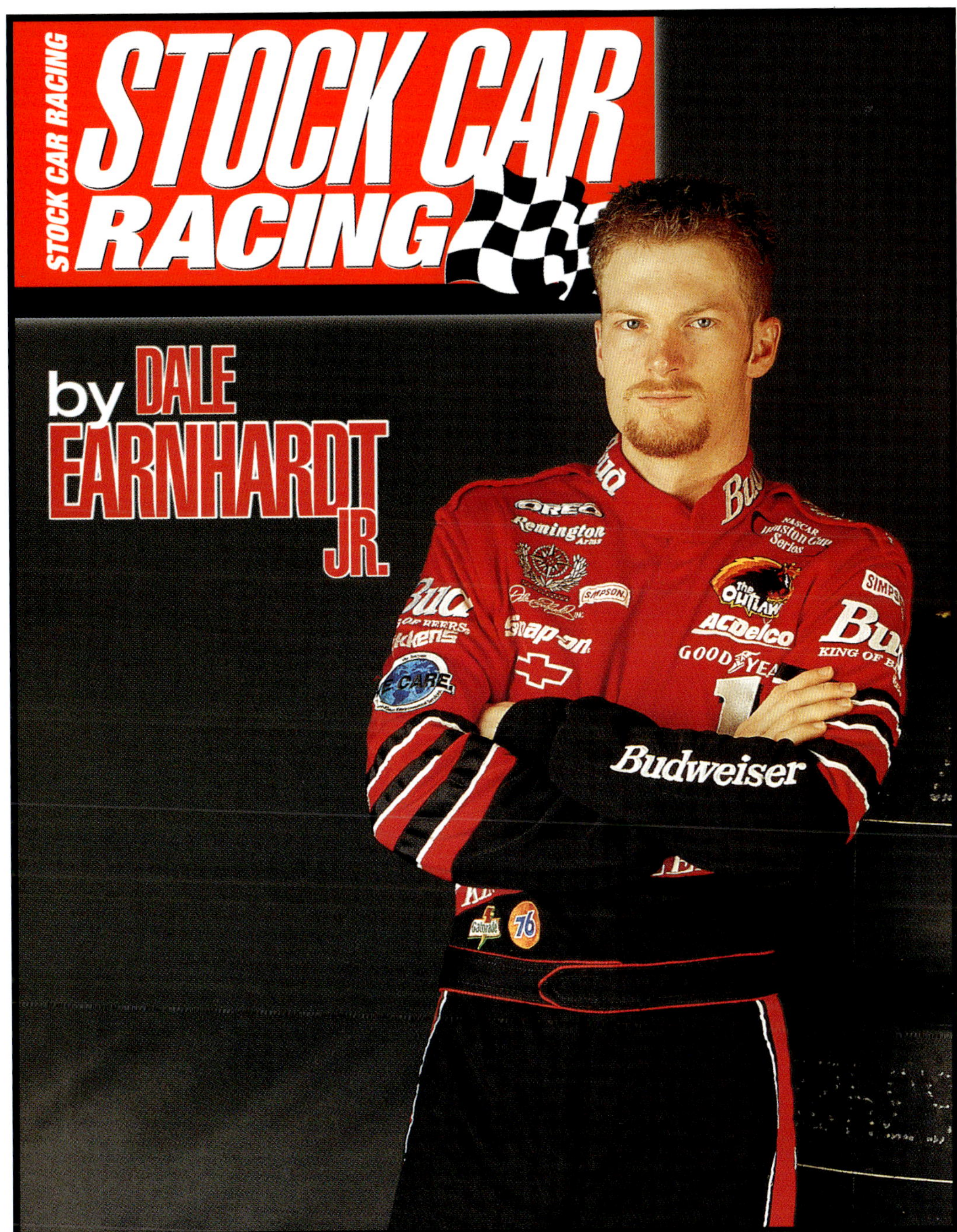

Earnhardt Jr.'s on-track success is the result of talent inherited from his father and the top-notch team that surrounds him. *Doug Miller*

Drafting

I get asked a lot about what it's like driving a car at 200 miles per hour. In some ways, it's easy. It's just simple physics. You have four tires on the car that connect to the asphalt or concrete. The key is to drive the car to the absolute edge of the tire's ability to grip that surface. It feels like you are on the verge of crashing at any moment when you're fighting a poor-handling car, but you just have to

"When I strap myself into my race car, the concentration level goes through the roof, especially at mentally draining tracks like Talladega,"—Earnhardt Jr. *Nigel Kinrade*

keep your foot in it and try to find that spot of ultimate speed.

At places like Talladega, I could teach just about anybody to go out there and run at top speed. It's so wide and the banking is so high, it's no big deal to drive at those speeds. I can have entire conversations on the radio with the team while I'm going down the endless back straight there. It seems like it's a mile long.

Now racing against 42 other cars at that speed is a different story. I don't think there is any other track that requires the concentration that Talladega does. It's not really because of the speed, but because we all run so close together with the NASCAR superspeedway aero rules. You have to be totally focused at all times when you're running cars three-wide. You almost hold your breath when you're in the middle line. You run a few laps like that and you can't believe nobody's crashed. I love it though. It's mentally challenging to run like that for three or four hours. It's like a chess match: if you get a good run, you may pass 10 cars, but you can drop back just as quickly if you are not totally focused on everything that goes on around you.

Some people said my dad could see the air coming around those cars, but I think he was able to use all of his knowledge about the air and the way to use it to pass guys. His peripheral vision was awesome—he could damn near see behind him so he was always one move ahead of the next guy.

The best way to describe the effects of the draft at Daytona or Talladega is to compare it to driving on the highway behind a big 18-wheeler. You swing your car into the passing lane to go around, and maybe for a split second the air coming off the truck makes your car feel light and jittery like it wants to be blown off the highway. Well, take that feeling and multiply your highway speed by three, and that's what it feels like all the time in the draft. The air does strange things—it sometimes feels like there's a beach ball in between you and the car in front of you as you close in. The air buffets your car and you can feel it vibrate and make your car move as you close up on his rear bumper.

"When you're running well at any track, your brain is totally focused and completely in tune with the race car and the racetrack,"—Earnhardt Jr. *Nigel Kinrade*

"Everyone on the team is focused on getting me to the front. Every crewman must be in sync during pit stops." – Earnhardt Jr. *Harold Hinson*

"When you do it enough, it becomes almost instinctive to be going that fast."—Earnhardt Jr.
Doug Miller

On the Inside

At nearly every other track, the four contact patches of those tires are absolutely critical and that's why you hear drivers moaning or cheering about a new tire compound. If you have a winning car setup and then Goodyear changes those tire compounds, you have to start all over again.

I was vocal a couple of times last year when I complained about the consistency of the tires, but I have to say that the tires this year have been great—very consistent from set to set and track to track. You get so in tune that you can feel every part of the car and each of those four tires as they go around the track. You are strapped in the seat so tight that you feel like the car is a part of you, and the entire snarling beast is rotating just about where your ass meets the seat. When it's right, it's a great feeling.

When you're running well at any track, your brain is totally focused and completely in tune with the race car and the racetrack. However, when you're running poorly, it can be a struggle. If you crashed early and you come back out on the track, you're 100 laps down and all you are doing is logging laps, whoa—that's tough! A lot of times my mind will wander to what is on my shopping list or what home improvements I need to make. Sometimes I think about the cold beer I can drink once I get home. You're like "get this over with, NOW!" I think that's also about the only time I really notice how hot it gets in the cars. I think on a hot day, it can be up to 140 degrees (Fahrenheit) in there, but I don't really notice unless we're under a yellow flag or the car really sucks.

"You can feel every part of the car, and each of those four tires on the track,"—Earnhardt Jr. *Harold Hinson*

Concentration and focus can be especially tough at Charlotte during the Coca-Cola 600, the longest race of the year. The last 100 miles are unbelievable because you are so drained, physically and mentally. You realize your head has been just lying on the headrest for lap after lap, and your eyes get blurry and out of focus. It's a weird feeling, but you just run out of concentration and energy.

Another twist in that race is that it finishes at night. I love racing at night—especially since I seem to do better. I won at Richmond and won The Winston last year under the lights, so people tease me that because I sleep a lot, I'm better at night. The lights make the cars look so cool, and it makes 'em look faster to the fans. However, to the drivers, it seems easier to concentrate under the lights. You can see the track better because all of the outside distractions like the fans in the stands or vehicles in the infield are now in the dark. It allows you to concentrate more on the track ahead of you than be distracted by all of the other things around you.

RADIO E

BY DALE EARNHARDT JR.

In a typical race, we will make adjustments to the car on nearly every pit stop. These include altering tire pressures, adding or removing spring rubbers, and making changes to the wedge in the rear of the car.

All of the changes are intended to get us up front and in place to challenge for the lead at the end of each race. These changes are critical, so we have to talk a lot on the radio throughout each race.

Mostly I talk with Tony Eury Jr., my cousin and the team's car chief. He is in charge of deciding what changes to make to the car before and during the race. I may also speak with Tony Eury Sr. (my uncle and the team's crew chief) or other team members as well. A lot of times we talk during yellow flag periods to ease the stress and the tension. Here are some fun (and not so fun) examples of our radio chatter during some recent races.

Sometimes Tony Jr. and his dad become more like cheerleaders than anything else. Here is one of our exchanges at Texas this year.

Tony Sr.: *"Save all the gas you can here. Don't use that accelerator pedal unless you have to."*

Me: *"Awwww. Are you kidding me? Jeez."*

Tony Sr.: *"We're still figuring, but we think you can make it. The 1 [Steve Park] and the 88 [Dale Jarrett] will need to stop again [for fuel]."*

Me: *"Well, I need to know here. If I go out there and bust my ass just trying to save fuel, I'm telling ya, it ain't gonna be driver error!"*

Tony Sr.: *"We just figured it. You'll have .87 of a gallon left as you get the checkered flag."*

Tony Jr.: *"Remember, this is Texas, you'll run out after you win this thing. Go for it. You got enough."*

I talk with other members of the crew on topics like engine temperatures. Here was a conversation I had with Jeff Clark, my team's engine tuner, who wanted to know about oil and water temperatures.

Clark: *"How do the gauges look?"*

Me: *"Nice. They're silver and they all have nice little red needles."*

I also had a little fun early into this year's running of the Coca-Cola 600 at Charlotte.

Me: *"Wassup, guys?! You guys have fun tonight, all right? ... Whoa! Somebody just threw a beer bottle at the car."*

Ty Norris (team spotter): *"Yeah, NASCAR is talking about it too."*

Me: *"It musta been a Sterling Marlin fan that threw it cuz it was a damn Coors Light."*

You never know when crap will happen. Things happen that make me so mad I want to chew through the steering wheel, but it's important to try and keep a positive attitude about coming back from adversity. We ran out of fuel at Fontana, lost a lap, and yet still came back to finish third. Here's what I had to say after losing that lap.

Me: *"Excellent. Excellent. This is JUST where I wanted to be. How could you guys let the car run outta gas?"*

Tony Jr.: *"Both computers figured we could go at least two more laps on fuel. Maybe it was because you were going so much faster."*

Me: *"I've just been driving so hard, working so damn hard, and now we're at the back."*

Tony Jr.: *"This whole damn weekend has been hard for all of us. We're just gonna get it back. Let's go get 'em and not let it end this way. You just gotta have that same attitude as last week. 'Ya gotta WANT to get back up front."*

"I'm in frequent radio contact with crew chief Tony Eury Sr. and others on the team, trying to figure out what adjustments to make during pit stops."—Earnhardt Jr. *Jeff Huneycutt*

"There's nothing like the feeling you get when your car is running well,"—Earnhardt Jr. *Sam Sharpe*

What I See

People ask if I can see things like fans in the grandstands while I'm driving that fast. I really can. When you run those speeds all the time, you just get used to them and everything seems to slow down. It must be like a professional baseball player who can almost see the stitching on a baseball that's been thrown at the plate at 95 miles per hour. The best moment

“You have to be totally focused at all times when you’re running cars three-wide,”—Earnhardt Jr. *Autostock, Webb*

Earnhardt Jr. has his father’s knack for getting around Talladega. *Nigel Kinrade*

"I don't think there is any other track that requires the concentration that Talladega does. It's not really because of the speed, but because we all run so close together with the NASCAR superspeedway aero rules,"—Earnhardt Jr. *Nigel Kinrade*

for that was when I won my first Winston Cup race at Texas. On the last lap, I was trying to concentrate but I could see the fans going crazy, waving and yelling and taking hundreds of flash photographs.

It's the same way talking to my crew. I guess some drivers only want to talk on the radio when they're on the straightaways. It is easier to do that, but you just get accustomed to doing things like talking to the crew while you're racing around. When it gets down to the last 10 laps though, everyone shuts up and just gives me how many laps are left. I need every ounce of my focus to go into winning that race.

The Winston last year was a good example. I had to pass 10 cars in eight laps, and my car was haulin' ass. So, the spotter and the crew just let me know how many laps were left, and I was able to drive around everyone. Believe me, I could see and hear the crowd on the last lap that night!

When you do it enough, it becomes almost instinctive to be going that fast. Sometimes it is much harder to slow down and come into the pits. It's harder than it looks to hit the right pit-road speed and then judge the distance in and out of the pit stall. Every tiny portion of a second can be so critical in the pits, and then you've got your crew guys and pit-crew guys from the teams ahead and behind you on pit lane zooming out in front of you. Then, you have to stay focused on making a good exit from the pit, so believe me, it's no 15-second vacation in there.

In the end, the technology of NASCAR is pretty simple. Everyone can get the best parts for their cars, and the cars are damn near equal. What wins races and championships are the people. The ability of a driver to communicate with his team and with his crew chief and car chief is critical. I learned that last year when we were struggling, and we've been much better this year because our people are working better together. It's so important that everyone is going in the same direction. If even one person is out of sync, you have almost no chance to win. When it does all go right, the feeling of victory is unbelievable. It is like a total release of energy and joy.

It's unbelievable! I look forward to my next chance to experience that feeling again. Soon.

"Concentration and focus can be especially tough at Charlotte during the Coca-Cola 600, the longest race of the year,"—Earnhardt Jr. *Nigel Kinrade*

"The four contact patches of those tires are absolutely critical and that's why you hear all of us drivers moaning or cheering about a new tire compound,"—Earnhardt Jr. *Nigel Kinrade*

8

JUNIOR, ON THE MOVE

BY LARRY COTHREN

From *Stock Car Racing* magazine, June 2002

NASCAR's Emerging Superstar Seeks Balance—and Victories—in His Life

In terms of clout and exposure in the marketplace, no driver in NASCAR history did as much as fast as Dale Earnhardt Jr. As the sport began to expand into territory previously outside the scope of NASCAR, it was Earnhardt Jr., more than any other driver, who was leading the way. Fame and fortune made it necessary to find harmony between his personal life and his professional life, however, as he lifted the sport on his young shoulders—while striving to perform on the track.

It's the day after Christmas and Dale Earnhardt Jr. and three friends embark on a road trip—one of those journeys you take when you're 27, wealthy, single, and anxious to kick around for a few days.

Another friend, a 21-year-old aspiring racer, is ready to make a career move by relocating from Buffalo to the Charlotte, North Carolina, area, and Junior and company head north in a red truck borrowed from Dale Earnhardt Chevrolet. They'll lend a hand to their New York buddy and live it up a little before they return to North Carolina.

"We got just past Greensboro [North Carolina] and turned off the interstate and just followed the compass from there on out, never got on another four-lane road," says Earnhardt. "It was pretty cool going through all those back roads. We went to Washington, D.C., and took our picture in front of the White House and in front of several of the monuments. We went through Gettysburg, Pennsylvania, and saw the battlefields, but it

Earnhardt Jr. and Jeff Gordon have become NASCAR's top drawing cards. *Nigel Kinrade*

Earnhardt Jr. at Daytona. *Harold Hinson*

Sponsors looking for maximum exposure in a limited amount of time come to Earnhardt Jr. *Harold Hinson*

was about two in the morning so it was kind of hard to see what was going on. We had a good time, just kind of messed around. It took us 22 hours to get up there because we were fiddling around all day and all night."

Road trips have become prized therapy for Earnhardt, mainly because they're not at all complicated. He and some buddies will simply hop into a car and drive somewhere. Usually they'll head to a place that doesn't have a local racetrack and—here's the key—where the locals don't follow racing. It's not so much what the trips offer as what they don't offer: no public appearances

A company usually gets a lot of bang for its buck as one of Earnhardt Jr.'s sponsors. *Harold Hinson*

to worry about, no strict schedules to adhere to, no one in your face all the time wanting you to do this or that or go there. "It's been more fun and beneficial for me when I do get time off to get totally away from it, to just go somewhere and get out of my element altogether," says Earnhardt.

So on the way back from Buffalo the week after Christmas, Earnhardt and his buddies detour a little west, going through Ohio. Not exactly a place to get away from racing, given the 50 or so racetracks that dot the Ohio landscape, but a diversion nonetheless.

"We stopped at a bar and spent the night and had a good time," Earnhardt says. "It was a lot of fun. It was kind of cool just to get away and be normal for a while. That was definitely a reality check to get you pumped up about the year and get you back into life."

Eye of the Storm

Taking a road trip and relaxing is one of those requisite life functions when you're suddenly the face of an entire sport and the whole world wants a piece of your time. Or when your new book, *Driver #8*, is on *The New York Times* bestseller list.

If we peep into Earnhardt's life for a glimpse of how hectic and demanding it is to always be in the public eye, then it's easy to see why something so innocent as a road trip, nothing more than an escape into a life resembling normalcy, is so valuable.

"The pressure is on him to continue to dazzle people with his image, his youth image, but he's got to deliver on the racetrack to keep that flame burning as strong as it is."
—Humpy Wheeler, track promoter

Earnhardt's sister, Kelley, and his publicist, Jade Gurss, are charged with the task of making sense and order of Earnhardt's professional life.

"It's like a big, giant jigsaw puzzle," says Gurss, who co-authored *Driver #8*. "You have a lot of pieces that are

Earnhardt Jr.'s on-track performance will be the ultimate gauge of his success. *Sam Sharpe*

There were glitches along the way, but Earnhardt Jr. claimed two Busch Series titles and five Winston Cup wins during his first four complete seasons in NASCAR. *Harold Hinson*

strewn all over the place and the challenge is to get them to fit in the best way possible. If he were to do all of the interviews or all of the appearances that are requested, he would be working 24 hours, 7 days a week. In the 10 days at Daytona, he did almost 95 interviews and nearly a dozen different sponsor appearances or functions, plus one massive autograph session for his new book."

This whirlwind of attention has surrounded Earnhardt since he took the Busch Series by storm and won back-to-back titles in 1998 and 1999. Since moving up to Winston Cup in 2000, when he promptly won two races and two poles, the intensity has increased as Earnhardt has almost single-handedly redefined stock car stardom, standing before people in places previously outside the NASCAR reach.

"We've been in a unique situation to go to different places and show people a little bit about NASCAR," says Earnhardt. "I might not represent the average mold for a NASCAR driver, but we were able to go to certain areas and [see] certain people throughout the country, especially toward the West Coast last year, with some of the articles and some of the networks we worked with, and showed ourselves and our sport to some interesting groups of people."

"You can't fool the public. When you dress somebody up and they do the dance, I think the public can tell the difference between somebody who's sincere and somebody who's not,"—Earnhardt Jr. *Harold Hinson*

Bringing Them In

It may seem odd that Dale Earnhardt Jr., with five wins in his first two Winston Cup seasons and a driver defined more by what he hasn't done on the track than what he has done, is the guy leading the NASCAR charge into new and mostly uncharted territory. It hasn't been that long since he was running late-models at Myrtle Beach, South Carolina. Now he's the face of the sport. Does that sound right?

Nearly all of the sporting press, and some on the outside looking in, think so. Look around and the guy is everywhere, including on magazine covers, in profiles, and everywhere a story can be told. Not to mention his appearances on MTV and *The Tonight Show* with Jay Leno, stories in *People* and *Rolling Stone*, and an interview in *Playboy*. This is definitely not his daddy's NASCAR, and that's the whole point.

Could it be that Dale Earnhardt Jr., the man of the hour in U.S. motorsports, is just the guy to stand before an entire country and represent stock car racing? He appears to meet all of the requirements.

"There are all kinds of things that I want to do, and I know that to be able to do [them] I'll have to continue as a race car driver and be good at it,"—Earnhardt Jr. *Nigel Kinrade*

NASCAR has long been about lineage, heritage, and many of the things in society that speak family—from American-made automobiles to children hoisted high in victory lane. NASCAR remains to this day a family-run business, one born over 50 years ago in the South and raised on traditional Southern values of hard work, loyalty, and dedication. So who better to represent the sport than a third-generation driver who is the son of the man who personified the very soul of stock car racing?

If Earnhardt Jr.'s bloodline alone is not enough to justify his place in the sport, consider this: the NASCAR growth curve demanded a younger audience, some maintain, and Earnhardt Jr. has met and exceeded the demand, and that alone has set him apart from those who've gone before him.

Humpy Wheeler, the Speedway Motorsports president who is a respected observer of the sport, says the timing was right for Earnhardt to make his mark.

"I think the pop culture that they introduced him to—or he introduced them to, that's debatable there—that's the first time that has happened, and I think it signifies the broader market we have today compared to even five years ago, particularly the youth market," says Wheeler.

"A decade ago we were beginning to be concerned that our demographics were reaching up too high as far as age was concerned. But Jeff Gordon, as he started winning, began to change that. Now Gordon is 30 and along comes Earnhardt Jr. at the right time with the right kind of personality. He's bringing that youth group along. It's just extraordinary how the demographics of this whole business are changing even as we speak."

It's not just teenagers and twenty-somethings who count themselves among the Earnhardt Jr. legion, though. Watch nearly an entire grandstand rise to its feet whenever Junior turns in a hot qualifying lap or makes a charge into the lead of a race, and you'll see a cross section of fans. He can bring that young person into the NASCAR fold, but he'll also grab the loyalty of that young person's father and grandfather. Some of those fans come from his father, no doubt, but they're in the fold either way.

A media swarm usually accompanies Earnhardt Jr. at the track. *Harold Hinson*

Another factor in Junior's ascension to the top of the sport is his ability to be himself. It's an old battle cry, but it remains a relevant one: a sport built on colorful characters lacks color today. And NASCAR drivers have long been poked fun at for their willingness to walk the walk and talk the talk to appease image-conscious sponsors. Then along comes Earnhardt Jr. with his laid-back personality, dash of color, and air of genuineness.

"I've been hearing a lot of talk about representing the sport and whether I would be one of those guys to do that or not, and maybe that's why I pressure myself on wearing just normal clothes every day and being myself," says Earnhardt. "I get a lot of flack from my sponsor for not having their logo all over my back and my shoulders and my head. Walking around in a pair of Adidas is pretty cool to me, and I want people to know that's who I am. And if you don't mind that guy who represents the sport wearing Adidas and the hat backwards, then that's fine. But if you do mind, then look somewhere else.

"You can't fool the public. When you dress somebody up and they do the dance, I think the public can tell the difference between somebody who's sincere and somebody who's not. Although I'm very proud of my relationship with all my sponsors—whether it be Budweiser or Drakkar or whoever—I don't feel like it's that necessary to don their logos everywhere I go. For some reason I think we can go further just being ourselves, and I think people will be more interested in that than billboarding."

A whirlwind of attention has surrounded "Little E" since he took the Busch Series by storm and won back-to-back titles in 1998 and 1999. *Harold Hinson*

Since moving up to Winston Cup in 2000, when he promptly won two races and two poles, the intensity has increased as Earnhardt Jr. has almost single-handedly redefined stock car stardom. *Harold Hinson*

HIT MAKER

We've seen the television commercial where Junior has problems with the CD player in his race car. But what artists would he choose if he did listen to CDs while driving 180 miles per hour? Here's what he told us, listed in the order he gave them:

1. **Bob Seger Greatest Hits**
2. **Foo Fighters**
3. **Matthew Good Band**
4. **Dr. Dre**
5. **Faith Hill**

Seeking Balance

At 27 years old, Earnhardt Jr. is firmly established in the sport at the same age his father was as a rookie. Junior epitomizes NASCAR cool while his father at 27 was a throwback to the sport's rough-hewn past. Where Earnhardt Sr.'s trademark was a bushy mustache, Junior's trademark is a ball cap worn backwards. Yet there's a mellow, somewhat domesticated nature that reveals itself, despite the road trips—or rather because of the road trips and the balance they lend to his life. Last year he even shut down "Club E," the basement nightclub at his house. No, he's not married yet, although he has said a wife and a son would go a long way toward leading a full life. But there's only so much you can pack into a life like the one he's lived the last few years.

Earnhardt Jr.'s time at the track is defined by a crush of reporters, microphones, cameras, and tape recorders.
Harold Hinson

"For the most part, it's been a lot of fun," says Earnhardt. "Even the really heavy structure, when it gets to be like we're doing something every day, it's still fun. There was a point last year and the year before that, every once in a while, where you seem to kind of lose touch with your home life and your family and your friends. This year I'm going to try to be a little more related toward that end of it, to be more involved with my friends and family than I was last year and spend some more time around them and try to get what I want out of my professional life and social life. We've been so busy running around and everything over the last couple of years that it's been really hard to maintain good relationships with my family and my friends. I'll try to do a little better with that this year.

"My mother moved from Norfolk [Virginia] about a year ago. I can just drive two miles down the road and see her, and that's really great. My sister [Kelley] had Carson, my little niece,

"If he were to do all the interviews or all of the appearances that are requested, he would be working 24 hours, 7 days a week,"—Jade Gurss, Earnhardt Jr.'s publicist. *Nigel Kinrade*

For over two decades Dale Earnhardt Sr. helped raise the popularity of stock car racing. Today, his son is at the forefront of the sport. *Harold Hinson*

Earnhardt Jr. teamed with car owner Richard Childress to win the season-opening Busch race at Daytona in 2002. *Nigel Kinrade*

about a year and a half ago, and that's been a lot of fun to be around her. Those things like that have made a lot of difference."

The specter of who he is and where he's going in his chosen profession is never far away, though. Ultimately, he realizes success will be gauged solely by his performance on the track. "There are all kinds of things that I want to do, and I know that to be able to do [them], I'll have to continue as a race car driver and be good at it," he says.

There are also the inevitable questions about goals and motivation that young, rich athletes face. Earnhardt fields questions about the importance of a breakout season with reference to a couple of drivers who reached the sport's pinnacle. "It's pretty important to me," he says. "I would say it's as important as it was to my father. If there's a gauge to go off of, I believe it would be just as important to me as it was to him or to Jeff Gordon or anybody else."

Wheeler has developed a theory on stardom and its place in NASCAR. During his five decades of involvement in stock car racing and from observing sports in general, he maintains that NASCAR got to where it is today by consistently having two superstars. The same formula applies to other sports, according to Wheeler. Since Dale Earnhardt Sr.'s fatal crash a year and a

Earnhardt Jr. and Teresa Earnhardt celebrated with Richard Childress and his wife, Judy, after the Daytona victory. *Nigel Kinrade*

half ago, Jeff Gordon stands as the sport's only true superstar, Wheeler maintains, and Earnhardt Jr. will have to win more often to claim a spot alongside Gordon.

"In this type of racing, to really be a superstar you've got to be a prolific winner," Wheeler says. "You can act like a superstar, you can look like a superstar, and just that charisma, those elements right there, will vault you up where everybody can see you. But you can't stay there unless you win races. So the pressure is on him to continue to dazzle people with his image, his youth image, but he's got to deliver on the racetrack to keep that flame burning as strong as it is.

"If he wins two or three races a year and we have the type of parity we have now, he might just be able to keep that going and might emerge as that second superstar."

For now, Earnhardt says his spot at the top of the sport, the spot that holds superstardom, will have to wait.

The media crush around Earnhardt Jr. at the track is exceeded only by the hordes of people lined up at his souvenir trailer. *Nigel Kinrade*

"I think," says Earnhardt, "that I will establish that a little more once we win some more races and maybe get closer to winning a championship. Right now we're just kind of growing."

Yet if we look out on the NASCAR horizon, we'll spot Earnhardt Jr. He'll be easy to pick out, and not just because of his hat turned backwards. He'll be the one with the road trips to make, family to embrace, friendships to nurture, and races to win. The key to being a superstar, after all, might be the ability to live with superstardom.

9

MEASURING UP

BY LARRY COTHREN

"The Only Thing That Matters to Me Is Driving and Kicking Butt."

At the end of 2002, with two seasons in the Busch Series and three in Winston Cup behind him, Earnhardt Jr. had seven Cup victories, a pair of Busch titles, and a solid place in the sport his father and grandfather helped build.

Earnhardt Jr. and the Budweiser crew won the fall race at Talladega in 2001 and both races there in 2002.
Harold Hinson

Earnhardt Jr. won at Texas and Richmond during his rookie season in Winston Cup—tying Davey Allison's record with 2 wins in his first 16 races—and he became the first rookie to win The Winston, NASCAR's annual all-star race. Then he won at Daytona in 2001, less than five months after his father's death in the Daytona 500. He also won at Dover's tough 1-mile track in his second Cup season; then, from the fall of his second season through 2002, he won three straight races at Talladega Superspeedway. Finally, he made a return trip to his old Busch Series stomping grounds in 2002 and won two of the three races he entered.

Make no mistake, this third-generation driver has made a mark in stock car racing where it counts most: on the track.

In a sport increasingly marked by parity, with 18 different winners in 2002 and 19 in 2001, Earnhardt Jr.'s domination at Talladega, in particular, has helped him continue the Earnhardt legacy in the stock car world. That, he says, makes him most proud.

"It makes me feel like I'm putting in my part as far as carrying on the Earnhardt name," he says. "That's real important to me, that there's an Earnhardt out there to cheer for, doing good enough to cheer for. That makes me feel good. I just want to keep on racking up accomplishments so that when

Talladega has become a stronghold for Earnhardt Jr. *Nigel Kinrade*

it's all done I can sit down and say I was a good race car driver and I made my daddy proud and I made my daddy's fans proud. But again, to some people it might not be enough because of what my daddy accomplished, you know?"

Accomplishments on the track have sometimes taken a back seat to the distractions in the modern NASCAR world. Yet strip away the glamour and attention that comes with being a Winston Cup star today, and what occurs on the racetrack becomes the only true method of evaluation. You especially don't measure an Earnhardt by the number of personal appearances he makes or by the volume of endorsements he can wrangle from the marketplace. An Earnhardt is measured by his performance between green flag and checkered flag. There is no other standard.

That was established when Ralph Earnhardt laid the groundwork for the Earnhardt legacy five decades ago, becoming the gold standard by which *all* short-track racers of his day were measured. By sheer grit and determination, Dale Earnhardt took his talent and success far beyond that of a regional star in a Southern sport. His record of 7 Winston Cup titles and 76 victories measured up against any racer of his time, as he became arguably the greatest stock car racer who ever lived.

An October, 2002 victory at Talladega earned Earnhardt Jr. a Winston bonus of one $1,000,000. *Harold Hinson*

That's the stage Dale Earnhardt Jr. stepped onto when he decided to become a stock car racer. No driver before him entered the sport with as much expected of him, or with a legacy quite like the one that came with being an Earnhardt.

Dale Earnhardt Sr. won 10 races at Talladega, and Earnhardt Jr. has continued that success. *Nigel Kinrade*

"I watched Davey Allison and Kyle Petty, watched those guys as a kid," says Earnhardt Jr., "knowing well that I was going to be under the same circumstances at some point in time, to have to weather the storm, so to speak, when you didn't live up to the expectations that everyone set for you. I know that winning seven championships or even coming close to that is a long shot for anybody. I don't care who you are, who your daddy was, what kind of car you've got, or who you race for, it's a long shot. That's a dream career for anybody. People who win those three and four championships only come along every 5 or 10 years."

Comparisons to his father's talent and success are inevitable, of course, and some see similarities to Dale Sr. in the driving style of Dale Jr. "He still opens that door," says Buddy Baker, a 15-race winner during his NASCAR driving career and a respected observer of the sport, of Earnhardt Jr. "He's one of those kids who will go for it. I think that's what everyone likes about Dale Jr. That part of his father, the apple didn't roll far from the tree, as far as stepping up when it's time to go to battle to win. I mean, he's there."

The pressure that comes with being the son of a stock car racing icon will always be there, too, regardless of what Earnhardt Jr. accomplishes on the track. The bar will forever be raised higher

"The most important thing always comes back to communication. For some reason, I've witnessed it firsthand these past couple of years and really have been concentrating on trying to get better at talking and working things out with the crew,"—Earnhardt Jr. *Nigel Kinrade*

"I've seen the good days and bad days, and you can really tell a difference when communication is not there. It has a direct effect on how you run and how things work out for you the rest of the day,"—Earnhardt Jr. *Harold Hinson*

Earnhardt Jr. has a tight bond with the Eurys. Crew chief Tony Eury Sr. is Earnhardt Jr.'s uncle and car chief Tony Eury Jr. is his first cousin. *Kevin Thorne*

Earnhardt Jr. competed in this Richard Childress–owned Busch car at Charlotte in May of 2002. *Nigel Kinrade*

and higher, but it's something he has dealt with since he began racing.

"I had more pressure on me in late-models because I couldn't produce, couldn't really win races or be dominant, or get a handful of wins at any racetrack we went to," says Earnhardt Jr. "We would run pretty good. We would run in the top five pretty much everywhere we went, but we couldn't win any races. So I felt like I was wasting my daddy's time and money doing what I was doing. It was kind of like a 'Man, this is a lot of fun. I wonder how long it's going to last' kind of deal.

"Now I don't feel quite as much pressure. I do feel that very soon, in the next couple of years, people are going to start pointing out that I'm getting close to 30, and I'm going to get a lot of pressure to step it up or whatever. Somebody might say, 'you know, it's time for him to take the next step.' I've already read some of those articles on the Internet. That's going to be like, 'Golly, let a guy live,' you know what I mean?

"I think the older I get, people are going to continue to make a story out of something. Most of it's true and some of it is just not really relevant or impor-

Earnhardt Jr. campaigned this colorful Busch Series Chevrolet at Richmond in September of 2002 and won handily. *Nigel Kinrade*

tant. I don't know if that will be pressure or not. Whenever I get on the Internet and read something that says, 'Well, he's 28 years old and it's time for him to take the next step into maturity,' that feels like pressure to me, you know?"

Despite the opinions and high standards others might establish, there is a level of comfort and confidence off the track, especially in how Earnhardt Jr. looks at his career. That contrasts sharply with the lack of confidence he suffered early in his career. "Me personally, I'm real happy with how I've done," he says. "I don't feel like, 'Darn, I didn't accomplish this.' I feel real good.

Earnhardt Jr. had two Busch Series victories in three starts during 2002. *Harold Hinson*

Earnhardt Jr. is very popular among his fellow drivers. *Harold Hinson*

When I was a teenager and just starting to race the late-model series, I was really intimidated and I wasn't too impressed at all with my driving, at that point. And I didn't think that how I was running was warranting a Winston Cup career.

"Now that I'm in Winston Cup, as long as I can look back and remember my attitude when I

Bonds form between drivers during a Winston Cup tour that covers nearly 40 weekends. *Sam Sharpe*

came into late-model racing, when I started driving, then I can look at my career now and say, 'Man, I really accomplished more than I anticipated.' When you're 17 years old and sitting in victory lane at a race your dad just won, you dream about winning your own race one day. You just can't imagine . . . 'Man, that's a long road to get there.'

"So yeah, I'm real happy. I've won several races, I've done some interesting things as far as races we've won, the ways we've won, the circumstances around the races. I mean, those are great accomplishments for me personally. I look forward to just continuing, but I don't want to level off, you know? I want to get better and better and better. I expect to do more, winning championships and stuff like that down the road. But right now, the first three years, I'm pretty happy with how it's going."

The Long Road

The transformation of Dale Earnhardt Jr., late-model racer, to Dale Earnhardt Jr., NASCAR's

Teammates Steve Park, Earnhardt Jr., and Michael Waltrip will forever be linked as the men who formed the first three-driver combination at Dale Earnhardt, Inc. *Nigel Kinrade*

The Budweiser Chevrolet at Indianapolis.
Harold Hinson

hottest young driver, wasn't something that just happened. Dale Earnhardt could have put all three of his children in a top Busch Series ride at any time, but he chose the only way he knew to mold talent: by letting each of his children grow and develop slowly, and by letting them learn, on their own, the same lessons he learned.

Now, at age 28, with a decade of full-time racing behind him, from late-models to Winston Cup, Earnhardt Jr. has developed a sense of what it takes to be successful in the sport.

"The thing that separates drivers nowadays, since we're all about equally as good as each other, is how you use your head, your judgment calls, your mentality, your outlook toward the season, your effort, your presence to the team, what you bring to the table as far as morale and leadership, things like that," he says.

There is the perception, nonetheless, that Earnhardt Jr., millionaire bachelor and promoter of Budweiser and wild times, is living too fast and loose to be at peak performance on the track.

"My personal opinion of him," says Bill Wilburn, crew chief for Rusty Wallace, "is when he gets his personal life to the point where he is 100 percent focused on driving the race car and that's

The red number 8 Chevrolet is one of the most popular cars in Winston Cup. *Sam Sharpe*

all that matters in his life, I don't see him being much different than his old man, because he's shown flashes of brilliance on every kind of track there is out there. The only place I haven't seen him really flash big is on a road course, and I'm sure that's coming. It's just a matter of time.

"I'm not knocking his personal life because the kid can do whatever he wants. He's a young man and he's got the world by the tail. He's a twenty-something, single millionaire and everything in the world is at his disposal. Hey, who wouldn't take advantage of the circumstances and

Carrying on the Earnhardt legacy is important to Earnhardt Jr. *Harold Hinson*

situations that come his way? The biggest thing, though, is when he gets that stuff where he wants it and gets his mind made up to go racing 100 percent—where that's all he thinks and lives and dreams and eats—it's going to be tough to handle him. Period."

Earnhardt Jr. is aware that many observers, both inside and outside the sport, look at his personal life and perceive a lack of dedication toward becoming the best driver he can possibly be.

"I would say two years ago that was right," he says. "Two years ago I didn't understand the commitment. I didn't understand how determined and how driven you have to be. Devoted, I

The Budweiser Chevrolet is often among the fastest cars during qualifying runs.
Harold Hinson

A road course victory is one of the remaining milestones for Earnhardt Jr.
Nigel Kinrade

A pole run at Kansas City was one of two in 2002.
Nigel Kinrade

guess, would be a good word. I wanted to have fun. Driving race cars was fun. If you look back at the 1960s and 1970s, there were drivers out there who raced hard on and off the track, and I think I'm probably no different.

"These days everything benches on 12 guys in a boardroom, and I think a lot of people have forgotten what racers were like 20, 30 years ago. I'm pretty tame compared to how they were back then. Now is a different time as far as how devoted and how determined and how much time you have to dedicate to the sport to be a good driver.

"I think when I first started in my rookie season in Winston Cup, I didn't understand the commitment, and it was just because it was new. Winston Cup and Busch are two different things. In Busch, you race like hell on Saturday and on Sunday you hit the lake and meet everybody and talk about it. It was fun, and there wasn't a lot of repercussions and restrictions. In the Winston Cup Series, man, it's just cookie-cutter. You've got to be by the book. Don't step out of line.

The July 2001 victory at Daytona came just five months after the death of his father at the track. It was a triumphant event for Earnhardt Jr., DEI, and NASCAR. *Nigel Kinrade*

Look the hell out. That's kind of the deal now.

"I've gotten better this past year. You build up this reputation and I think it's really hard to dismantle it in the matter of a year. But I'm living quite the different lifestyle than I did two years ago."

One of his more notorious playgrounds was in the basement of his house, which he converted into a full-fledged private nightclub with a bar and dance floor. However, "Club E" was shut down in 2001.

"I put a pool table down there and I go down and play a little pool with three or four of my buddies," he says. "We don't have the old club we used to have. I remodeled my house, and as far as taking better care of it and as far as who comes in and who comes out, it's a little more strict. I don't have any visitors. I used to have people come over all the time, on Wednesday nights, on Sunday nights after the race. We used to gather up a group of 10 or 15 or so and just raise hell. I haven't done that in a long, long time it seems like.

"I remodeled my house, got some nice shrubbery up front, got my shop going in the back and everything. Bought some land, been working on my land. Done a lot of growing up in the last two years."

A Bit of Maturity

With seemingly the whole world watching over the past three seasons, particularly the last two since his father died, Earnhardt Jr. has faced the demands and challenges

The Budweiser crew helped Earnhardt Jr. to an 11th-place points finish in 2002. *Harold Hinson*

Over the past two years, Earnhardt Jr. has worked to change his reputation for fast living. *Nigel Kinrade*

of being a superstar. He has, clearly, learned the lessons outside the track his father sought to instill in him. When Earnhardt Jr. says he has matured, many who know him agree. Ernie Hammonds, the self-described "general flunky" from the Myrtle Beach days, says Earnhardt Jr. has matured quite well from the shy teenager who signed autographs for 50 people without ever raising his head.

"I was at an autograph session in Bristol [Tennessee], downtown on State Street, in August [2002]," says Hammonds, "and to each and every person who went through that line he said, 'Thank you very much. Have a good day.' *To each and every one that went through that line*. So he's come a long, long way. As a *person*.

"He had every opportunity in the world to act like a couple of the other drivers have acted, and probably every reason in the world to be a jerk, and hasn't been. He's changed some, but I

Earnhardt Jr. has become a fixture in the sport while driving for the team his father built. *Nigel Kinrade*

In five years, Earnhardt Jr. has become one of the most recognizable figures in NASCAR. *Kevin Thorne*

don't think it's fair for people to say his head has gotten big since he did this. If I had somebody poking at me every 30 seconds, I would probably be that way, and I wouldn't be as nice as he is."

Yet, through the grind of being constantly in the spotlight and being pulled in several different directions, there is a yearning for simpler times, says Earnhardt Jr.

"I miss going down to Myrtle Beach, taking a couple of checks, one for tires and one for pit passes," he says, "and having a great time, you know? These days you don't know whether you're coming or going or who you're getting ready to meet. You're meeting people. 'Hey, I'm going to introduce you to this guy. I'm going to introduce you to this guy.' And you're remembering their

Earnhardt Jr. uses the same number 8 that his grandfather, Ralph Earnhardt, used while winning hundreds of short-track races in the Carolinas. *Nigel Kinrade*

Earnhardt Jr. and teammate Michael Waltrip chat during the 2002 season. *Harold Hinson*

name and face, not because the meeting was interesting and you enjoyed their company, but because you don't know whether you're going to run into them down the road and you've got to remember who they were. You know what I mean? Everything is kind of not for the right reasons anymore.

"When I get in the car, that's the best time I can get in there, and I'm doing that because I like it. That's why I've always done it, but everything kind of changed. In a matter of two years, my outlook toward things has changed dramatically. It's amazing. I can't imagine what it's going to be like five years from now."

Has he changed in two years because of the death of his father?

"I don't know if that was a cornerstone or anything, but I used to get really, really nervous about talking in front of a big crowd," he says. "Now it's a breeze, because there are so many other things out there that are tougher and harder that you'll go through, and that's nothing. I don't know whether it's because I've gotten better at it or I just don't really care anymore about it. The only thing that matters to me is driving and kicking butt and being The Man out there."

The Budweiser Chevrolet has become a force to be reckoned with in the Winston Cup Series. *Nigel Kinrade*

DALE EARNHARDT JR.'S NASCAR APPEARANCES

WINSTON CUP (1999–2002)

Year	Race No.	Race	Finish	Start	Car	Laps	Condition	Winnings
2002	1	Daytona – Daytona 500	29	5	No. 8 Budweiser Chevrolet	171	Running	$175,137
	2	Rockingham – Subway 400	26	29	No. 8 Budweiser Chevrolet	391	Running	77,587
	3	Las Vegas – UAW-DaimlerChrysler 400	16	35	No. 8 Budweiser Chevrolet	267	Running	84,875
	4	Atlanta – MBNA America 500	2	3	No. 8 Budweiser Chevrolet	325	Running	104,600
	5	Darlington – Carolina Dodge Dealers 400	4	23	No. 8 Budweiser Chevrolet	293	Running	96,542
	6	Bristol – Food City 500	4	23	No. 8 Budweiser Chevrolet	500	Running	87,850
	7	Texas – Samsung/RadioShack 500	42	9	No. 8 Budweiser Chevrolet	192	Accident	85,722
	8	Martinsville – Virginia 500	5	3	No. 8 Budweiser Chevrolet	500	Running	88,862
	9	Talladega – Aaron's 499	1	4	No. 8 Budweiser Chevrolet	188	Running	184,830
	10	California – NAPA Auto Parts 500	36	9	No. 8 Budweiser Chevrolet	225	Accident	89,987
	11	Richmond – Pontiac Excitement 400	36	2	No. 8 Budweiser Chevrolet	327	Running	76,862
	12	Charlotte – Coca-Cola 600	35	6	No. 8 Budweiser Chevrolet	371	Engine	76,575
	13	Dover – MBNA Platinum 400	30	30	No. 8 Budweiser Chevrolet	396	Running	87,492
	14	Pocono – Pocono 500	12	14	No. 8 Budweiser Chevrolet	200	Running	78,612
	15	Michigan – Sirius Satellite Radio 400	22	3	No. 8 Budweiser Chevrolet	199	Running	80,677
	16	Sears Point – Dodge/Save Mart 350	30	23	No. 8 Budweiser Chevrolet	109	Running	84,607
	17	Daytona – Pepsi 400	6	9	No. 8 Budweiser Chevrolet	160	Running	105,537
	18	Chicagoland – Tropicana 400	10	9	No. 8 Budweiser Chevrolet	267	Running	102,437
	19	New Hampshire – New England 300	23	28	No. 8 Budweiser Chevrolet	300	Running	64,925
	20	Pocono – Pennsylvania 500	37	11	No. 8 Budweiser Chevrolet	144	Running	55,100
	21	Indianapolis – Brickyard 400	22	3	No. 8 Budweiser Chevrolet	160	Running	125,280
	22	Watkins Glen – Sirius Satellite Radio at The Glen	35	21	No. 8 Budweiser Chevrolet	84	Running	57,605
	23	Michigan – Pepsi 400	10	1	No. 8 Budweiser Chevrolet	200	Running	78,740
	24	Bristol – Sharpie 500	3	2	No. 8 Budweiser Chevrolet	500	Running	131,090
	25	Darlington – Mountain Dew Southern 500	16	15	No. 8 Budweiser Chevrolet	367	Running	67,185
	26	Richmond – Chevy Monte Carlo 400	4	24	No. 8 Budweiser Chevrolet	400	Running	87,205
	27	New Hampshire – New Hampshire 300	11	37	No. 8 Budweiser Chevrolet	207	Running	70,425
	28	Dover – MBNA All-American Heroes 400	24	3	No. 8 Budweiser Chevrolet	396	Running	66,240
	29	Kansas – Protection One 400	6	1	No. 8 Budweiser Chevrolet	267	Running	96,700

Year	Race No.	Race	Finish	Start	Car	Laps	Condition	Winnings
	30	Talladega – EA Sports 500	1	13	No. 8 Budweiser Chevrolet	188	Running	$166,040
	31	Charlotte – UAW-GM Quality 500	9	13	No. 8 Budweiser Chevrolet	334	Running	69,300
	32	Martinsville – Old Dominion 500	4	10	No. 8 Budweiser Chevrolet	500	Running	75,870
	33	Atlanta – NAPA 500	5	13	No. 8 Budweiser Chevrolet	248	Running	97,950
	34	Rockingham – Pop Secret 400	34	22	No. 8 Budweiser Chevrolet	389	Running	59,100
	35	Phoenix – Checker Auto Parts 500	5	3	No. 8 Budweiser Chevrolet	312	Running	97,850
	36	Homestead – Ford 400	21	2	No. 8 Budweiser Chevrolet	267	Running	60,825
2001	1	Daytona – Daytona 500	2	6	No. 8 Budweiser Chevrolet	200	Running	975,907
	2	Rockingham – Dura Lube 400	43	25	No. 8 Budweiser Chevrolet	0	Accident	66,924
	3	Las Vegas – UAW-DaimlerChrysler 400	23	32	No. 8 Budweiser Chevrolet	266	Running	90,398
	4	Atlanta – Cracker Barrel 500	15	17	No. 8 Budweiser Chevrolet	324	Running	73,723
	5	Darlington – Carolina Dodge Dealers 400	34	16	No. 8 Budweiser Chevrolet	285	Running	64,138
	6	Bristol – Food City 500	31	9	No. 8 Budweiser Chevrolet	456	Running	77,288
	7	Texas – Harrah's 500	8	1	No. 8 Budweiser Chevrolet	334	Running	133,573
	8	Martinsville – Virginia 500	11	10	No. 8 Budweiser Chevrolet	500	Running	71,233
	9	Talladega – Talladega 500	8	19	No. 8 Budweiser Chevrolet	188	Running	90,178
	10	California – NAPA Auto Parts 500	3	38	No. 8 Budweiser Chevrolet	250	Running	136,873
	11	Richmond – Pontiac Excitement 400	7	14	No. 8 Budweiser Chevrolet	400	Running	75,348
	12	Charlotte – Coca-Cola 600	25	19	No. 8 Budweiser Chevrolet	398	Running	85,778
	13	Dover – MBNA Platinum 400	3	11	No. 8 Budweiser Chevrolet	400	Running	113,648
	14	Michigan – Kmart 400	39	7	No. 8 Budweiser Chevrolet	171	Engine	69,593
	15	Pocono – Pocono 500	20	6	No. 8 Budweiser Chevrolet	199	Running	73,288
	16	Sears Point – Dodge/Save Mart 350	19	37	No. 8 Budweiser Chevrolet	112	Running	75,718
	17	Daytona – Pepsi 400	1	13	No. 8 Budweiser Chevrolet	160	Running	185,873
	18	Chicagoland – Tropicana 400	11	36	No. 8 Budweiser Chevrolet	267	Running	70,800
	19	New Hampshire – New England 300	9	29	No. 8 Budweiser Chevrolet	300	Running	78,473
	20	Pocono – Pennsylvania 500	2	12	No. 8 Budweiser Chevrolet	200	Running	119,923
	21	Indianapolis – Brickyard 400	10	36	No. 8 Budweiser Chevrolet	160	Running	161,833
	22	Watkins Glen – Global Crossing at The Glen	12	27	No. 8 Budweiser Chevrolet	90	Running	69,788
	23	Michigan – Pepsi 400	12	12	No. 8 Budweiser Chevrolet	162	Running	75,188
	24	Bristol – Sharpie 500	14	24	No. 8 Budweiser Chevrolet	500	Running	85,063
	25	Darlington – Mountain Dew Southern 500	17	18	No. 8 Budweiser Chevrolet	367	Running	74,788
	26	Richmond – Monte Carlo 400	3	8	No. 8 Budweiser Chevrolet	400	Running	106,053
	27	Dover – MBNA Cal Ripken Jr. 400	1	3	No. 8 Budweiser Chevrolet	400	Running	168,858
	28	Kansas – Protection One 400	33	22	No. 8 Budweiser Chevrolet	228	Accident	80,423
	29	Charlotte – UAW-GM Quality 500	4	9	No. 8 Budweiser Chevrolet	334	Running	104,923
	30	Martinsville – Old Dominion 500	27	2	No. 8 Budweiser Chevrolet	496	Running	67,623
	31	Talladega – EA Sports 500	1	6	No. 8 Budweiser Chevrolet	188	Running	165,773
	32	Phoenix – Checker Auto Parts 500	37	30	No. 8 Budweiser Chevrolet	287	Engine	73,248
	33	Rockingham – Pop Secret 400	15	21	No. 8 Budweiser Chevrolet	391	Running	75,223
	34	Homestead – Pennzoil 400	15	17	No. 8 Budweiser Chevrolet	267	Running	77,373

Year	Race No.	Race	Finish	Start	Car	Laps	Condition	Winnings
	35	Atlanta – NAPA 500	7	1	No. 8 Budweiser Chevrolet	325	Running	$122,123
	36	New Hampshire – New Hampshire 300	24	8	No. 8 Budweiser Chevrolet	297	Running	72,948
2000	1	Daytona – Daytona 500	13	8	No. 8 Budweiser Chevrolet	200	Running	107,775
	2	Rockingham – DuraLube/Kmart 400	19	7	No. 8 Budweiser Chevrolet	390	Running	27,110
	3	Las Vegas – carsdirect.com 400	10	3	No. 8 Budweiser Chevrolet	148	Running	76,275
	4	Atlanta – Cracker Barrel 500	29	2	No. 8 Budweiser Chevrolet	307	Running	25,805
	5	Darlington – Mall.com 400	40	10	No. 8 Budweiser Chevrolet	213	Accident	23,375
	6	Bristol – Food City 500	38	12	No. 8 Budweiser Chevrolet	401	Accident	26,235
	7	Texas – DIRECTV 500	1	4	No. 8 Budweiser Chevrolet	334	Running	374,675
	8	Martinsville – Goody's Body Pain 500	26	22	No. 8 Budweiser Chevrolet	496	Running	33,975
	9	Talladega – Die Hard 500	42	6	No. 8 Budweiser Chevrolet	113	Engine	42,850
	10	California – NAPA Auto Parts 500	12	20	No. 8 Budweiser Chevrolet	250	Running	52,375
	11	Richmond – Pontiac Excitement 400	1	5	No. 8 Budweiser Chevrolet	400	Running	118,850
	12	Charlotte – Coca-Cola 600	4	1	No. 8 Budweiser Chevrolet	400	Running	110,900
	13	Dover – MBNA Platinum 400	10	6	No. 8 Budweiser Chevrolet	398	Running	54,785
	14	Michigan – Kmart 400	13	6	No. 8 Budweiser Chevrolet	193	Running	38,200
	15	Pocono – Pocono 500	19	15	No. 8 Budweiser Chevrolet	200	Running	44,015
	16	Sears Point – Save Mart/Kragen 350k	24	31	No. 8 Budweiser Chevrolet	112	Running	47,550
	17	Daytona – Pepsi 400	35	31	No. 8 Budweiser Chevrolet	159	Running	45,330
	18	New Hampshire – thatlook.com 300	21	26	No. 8 Budweiser Chevrolet	272	Running	51,075
	19	Pocono – Pennsylvania 500	13	15	No. 8 Budweiser Chevrolet	200	Running	43,015
	20	Indianapolis – Brickyard 400	13	6	No. 8 Budweiser Chevrolet	160	Running	124,260
	21	Watkins Glen – Global Crossing at The Glen	40	14	No. 8 Budweiser Chevrolet	60	Transmission	34,830
	22	Michigan – Pepsi 400	31	1	No. 8 Budweiser Chevrolet	199	Running	37,730
	23	Bristol – goracing.com 500	21	12	No. 8 Budweiser Chevrolet	499	Running	40,885
	24	Darlington – Pepsi Southern 500	11	13	No. 8 Budweiser Chevrolet	328	Running	46,345
	25	Richmond – Chevrolet Monte Carlo 400	13	31	No. 8 Budweiser Chevrolet	399	Running	37,905
	26	New Hampshire – Dura Lube 300	31	20	No. 8 Budweiser Chevrolet	288	Accident	47,100
	27	Dover – MBNA.com 400	16	13	No. 8 Budweiser Chevrolet	399	Running	55,190
	28	Martinsville – NAPA AutoCare 500	36	32	No. 8 Budweiser Chevrolet	406	Accident	31,500
	29	Charlotte – UAW-GM Quality 500	19	13	No. 8 Budweiser Chevrolet	334	Running	34,390
	30	Talladega – Winston 500	14	3	No. 8 Budweiser Chevrolet	188	Running	49,260
	31	Rockingham – Pop Secret 400	34	37	No. 8 Budweiser Chevrolet	301	Accident	33,550
	32	Phoenix – Checker/DuraLube 500K	27	18	No. 8 Budweiser Chevrolet	311	Running	49,625
	33	Homestead – Pennzoil 400	13	12	No. 8 Budweiser Chevrolet	265	Running	49,525
	34	Atlanta – NAPA 500	20	40	No. 8 Budweiser Chevrolet	322	Running	50,810
1999	12	Charlotte – Coca-Cola 600	16	8	No. 8 Budweiser Chevrolet	397	Running	36,250
	18	New Hampshire – Jiffy Lube 300	43	13	No. 8 Budweiser Chevrolet	44	Ignition	36,475
	22	Michigan – Pepsi 400	24	17	No. 8 Budweiser Chevrolet	199	Running	21,765
	25	Richmond – Exide Select Batteries 400	10	21	No. 8 Budweiser Chevrolet	399	Running	29,905
	34	Atlanta – NAPA 500	14	13	No. 8 Budweiser Chevrolet	324	Running	37,700

BUSCH SERIES (1997–2002)

Year	Race No.	Race	Finish	Start	Car	Laps	Condition	Winnings
2002	1	Daytona – EAS/GNC Live Well 300	1	4	No. 3 Nabisco Oreo Chevrolet	120	Running	$79,175
	13	Charlotte – CARQUEST Auto Parts 300	36	11	No. 3 Nabisco Nilla Chevrolet	83	Accident	10,950
	26	Richmond – Funai 250	1	1	No. 8 Warner Brothers Chevrolet	250	Running	19,850
2001	14	Charlotte – CARQUEST Auto Parts 300	29	33	No. 87 Cellular One Chevrolet	197	Running	9,920
1999	1	Daytona – NAPA Auto Parts 300	14	16	No. 3 ACDelco Chevrolet	120	Running	30,200
	2	Rockingham – ALLTEL 200	35	1	No. 3 ACDelco Chevrolet	126	Accident	15,285
	3	Las Vegas – Sam's Town 300	6	13	No. 3 ACDelco Chevrolet	200	Running	36,400
	4	Atlanta – Yellow Freight 300	3	15	No. 3 ACDelco Chevrolet	195	Running	29,000
	5	Darlington – Diamond Hill Plywood 200	11	5	No. 3 ACDelco Chevrolet	146	Running	17,035
	6	Texas – Coca-Cola 300	10	12	No. 3 ACDelco Chevrolet	163	Running	28,600
	7	Nashville – BellSouth Mobility 320	9	1	No. 3 ACDelco Chevrolet	320	Running	19,235
	8	Bristol – Moore's Snacks 250	2	7	No. 3 ACDelco Chevrolet	250	Running	27,910
	9	Talladega – Touchstone Energy 300	6	17	No. 3 ACDelco Chevrolet	113	Running	27,125
	10	California – Auto Club 300	3	1	No. 3 ACDelco Chevrolet	150	Running	53,875
	11	New Hampshire – New Hampshire 200	34	5	No. 3 ACDelco Chevrolet	193	Accident	17,740
	12	Richmond – Hardee's 250	32	4	No. 3 ACDelco Chevrolet	201	Rear End	15,540
	13	Nazareth – First Union 200	2	3	No. 3 ACDelco Chevrolet	168	Running	26,325
	14	Charlotte – CARQUEST Auto Parts 300	2	22	No. 3 ACDelco Chevrolet	200	Running	45,150
	15	Dover – MBNA Platinum 200	1	15	No. 3 ACDelco Chevrolet	200	Running	44,725
	16	South Boston – Textilease Medique 300	1	1	No. 3 ACDelco Chevrolet	300	Running	36,300
	17	Watkins Glen – Lysol 200	1	3	No. 3 ACDelco Chevrolet	82	Running	37,800
	18	Milwaukee – Die Hard 250	3	15	No. 3 ACDelco Chevrolet	250	Running	27,750
	19	Myrtle Beach – Myrtle Beach 250	25	1	No. 3 ACDelco Chevrolet	247	Running	18,780
	20	Pikes Peak – NAPA AutoCare 250	36	12	No. 3 ACDelco Chevrolet	239	Accident	19,350
	21	Gateway International – CARQUEST Auto Parts 250	1	18	No. 3 ACDelco Chevrolet	200	Running	51,775
	22	Indianapolis Raceway Park – Kroger 200	5	3	No. 3 ACDelco Chevrolet	200	Running	19,125
	23	Michigan – NAPA 200	1	3	No. 3 ACDelco Chevrolet	100	Running	43,650
	24	Bristol – Food City 250	3	5	No. 3 ACDelco Chevrolet	250	Running	25,615
	25	Darlington – Dura Lube 200	12	14	No. 3 ACDelco Chevrolet	146	Running	17,005
	26	Richmond – Autolite Platinum 250	1	20	No. 3 ACDelco Chevrolet	250	Running	35,950
	27	Dover – MBNA.com 200	33	3	No. 3 ACDelco Chevrolet	176	Running	17,300
	28	Charlotte – All Pro Bumper to Bumper 300	5	20	No. 3 ACDelco Chevrolet	200	Running	32,450
	29	Rockingham – Kmart 200	13	2	No. 3 ACDelco Chevrolet	196	Running	17,950
	30	Memphis – Sam's Town 250	2	8	No. 3 ACDelco Chevrolet	250	Running	40,475

Year	Race No.	Race	Finish	Start	Car	Laps	Condition	Winnings
	31	Phoenix – Outback Steakhouse 200	2	5	No. 3 ACDelco Chevrolet	200	Running	$41,075
	32	Homestead – HotWheels.com 300	2	8	No. 3 ACDelco Chevrolet	200	Running	68,700
1998	1	Daytona – NAPA Auto Parts 300	37	3	No. 3 ACDelco Chevrolet	81	Accident	22,925
	2	Rockingham – GM Goodwrench Service Plus 200	16	6	No. 3 ACDelco Chevrolet	197	Running	12,100
	3	Las Vegas – Sam's Town Las Vegas 300	2	8	No. 3 ACDelco Chevrolet	200	Running	59,000
	4	Nashville Speedway USA – BellSouth/Opryland 320	3	7	No. 3 ACDelco Chevrolet	320	Running	24,775
	5	Darlington – Diamond Hill Plywood 200	10	37	No. 3 ACDelco Chevrolet	147	Running	14,425
	6	Bristol – Moore's Snacks 250	2	1	No. 3 ACDelco Chevrolet	250	Running	21,260
	7	Texas – Coca-Cola 300	1	16	No. 3 ACDelco Chevrolet	200	Running	66,075
	8	Hickory – Galaxy Foods 300	8	16	No. 3 ACDelco Chevrolet	300	Running	13,045
	9	Talladega – Touchstone Energy 300	32	4	No. 3 ACDelco Chevrolet	43	Handling	14,300
	10	New Hampshire – Gumout Long Life Formula 200	10	16	No. 3 ACDelco Chevrolet	200	Running	13,475
	11	Nazareth – First Union 200	28	1	No. 3 ACDelco Chevrolet	188	Running	12,925
	12	Charlotte – CARQUEST Auto Parts 300	30	23	No. 3 ACDelco Chevrolet	175	Running	13,245
	13	Dover – MBNA Platinum 200	1	16	No. 3 ACDelco Chevrolet	200	Running	34,075
	14	Richmond – Hardee's 250	2	3	No. 3 ACDelco Chevrolet	250	Running	27,095
	15	Pikes Peak – Lycos.com 250	10	5	No. 3 ACDelco Chevrolet	250	Running	24,000
	16	Watkins Glen – Lysol 200	8	12	No. 3 ACDelco Chevrolet	82	Running	14,150
	17	Milwaukee – Die Hard 250	1	2	No. 3 ACDelco Chevrolet	250	Running	39,625
	18	Myrtle Beach – Myrtle Beach 250	5	3	No. 3 ACDelco Chevrolet	250	Running	17,075
	19	California – Kenwood Home & Car Audio 300	1	2	No. 3 ACDelco Chevrolet	150	Running	68,175
	20	South Boston – Lycos.com 300	13	1	No. 3 ACDelco Chevrolet	299	Running	16,925
	21	Indianapolis Raceway Park – Kroger 200	1	16	No. 3 ACDelco Chevrolet	200	Running	34,225
	22	Michigan – Pepsi 200	5	10	No. 3 ACDelco Chevrolet	100	Running	19,275
	23	Bristol – Food City 250	15	4	No. 3 ACDelco Chevrolet	246	Running	14,490
	24	Darlington – Dura Lube 200	2	34	No. 3 ACDelco Chevrolet	147	Running	28,375
	25	Richmond – Autolite Platinum 250	1	2	No. 3 ACDelco Chevrolet	250	Running	35,600
	26	Dover – MBNA Gold 200	8	17	No. 3 ACDelco Chevrolet	199	Running	16,590
	27	Charlotte – All Pro Bumper to Bumper 300	3	5	No. 3 ACDelco Chevrolet	200	Running	39,325
	28	Gateway International – CARQUEST Auto Parts 250	1	13	No. 3 ACDelco Chevrolet	200	Running	48,525
	29	Rockingham – ACDelco 200	14	8	No. 3 ACDelco Chevrolet	197	Running	15,095
	30	Atlanta – Stihl Outdoor Power Tools 300	2	6	No. 3 ACDelco Chevrolet	195	Running	37,925
	31	Homestead – Jiffy Lube Miami 300	42	15	No. 3 ACDelco Chevrolet	89	Engine	17,470
1997	10	Nashville Speedway USA – BellSouth/Opryland 320	39	19	No. 31 Gargoyles Chevrolet	93	Oil Pump	2,665
	17	Watkins Glen – Lysol 200	39	9	No. 31 Gargoyles Chevrolet	12	Engine	2,525
	20	Gateway International – Gateway 300	38	18	No. 31 Gargoyles Chevrolet	61	Accident	11,400
	22	Michigan – Detroit Gasket 200	7	18	No. 31 Gargoyles Chevrolet	100	Running	6,295
	23	Bristol – Food City 250	22	2	No. 7 Gargoyles Chevrolet	247	Running	4,355
	28	California – Kenwood 300	34	11	No. 31 Gargoyles Chevrolet	129	Handling	11,150
	29	Rockingham – ACDelco 200	16	24	No. 31 ACDelco Chevrolet	197	Running	4,175
	30	Homestead – Jiffy Lube Miami 300	13	14	No. 31 Gargoyles Chevrolet	199	Running	10,315

INDEX